CW00350266

About the Author

Maynard Davies was born in Staffordshire. He left school at the earliest opportunity with no qualifications as he had dyslexia, a condition not recognised in his day. He arrived at Theodore Mountford's curing factory in search of a job as a bacon-curing apprentice. He was told to be at the factory at 6am every morning, and if he was a minute later, not to bother turning up.

This was the start of Maynard's life as an apprentice in one of the finest professions, that of traditional bacon curing and old-fashioned meat production. It eventually took him to many places including America and Europe. He met many interesting people and learnt the traditional methods of producing good food. He was hired as a consultant to many major food companies in England.

Maynard married and had four daughters, moving to the Peak District where he farmed, kept his own pig herd and produced his own bacon. When the girls grew up, he and his wife Trisha moved to Shropshire for the gentler climate and there they opened Maynard's Farm shop, working there until he sold up and retired in 2001. He has since been writing books (he is the author of *Maynard: Adventures of a Bacon Curer*, and *Maynard: Secrets of a Bacon Curer*) and acting as a food consultant. Sadly, Trisha died in 1990. He met his present wife Ann a few years later, and she now helps Maynard to put together his books. They live in north Shropshire.

Manual of a Traditional Bacon Curer

MAYNARD DAVIES

First published in Great Britain by Merlin Unwin Books Ltd, 2009
Reprinted in 2011

Published by:

Merlin Unwin Books Ltd
Palmers House
7 Corve Street
Ludlow
Shropshire SY8 1DB
U.K.

www.merlinunwin.co.uk

The author asserts his moral right to be identified with this work.

ISBN 978-1-906122-08-9

Typeset in 12 point Bembo by Merlin Unwin Books

Printed in China on behalf of Latitude Press Ltd

MANUAL OF A TRADITIONAL BACON CURER

MAYNARD DAVIES

with Ann Purchase

MERLIN UNWIN BOOKS

ACKNOWLEDGEMENTS

I would like to thank Theo Mountford for introducing me to the food industry and passing his extensive knowledge and recipes to me. I would also like to thank all the people who have opened their doors and extended their courtesy to me, including: Melvyn and Nita Ling of Appleyards, Neil Hollingsworth of Duke-shill, the Themans family of Wenlock Edge Farm, Sandy Boyd and the staff at the Ludlow Food Centre and Alan Ball of Bings Heath Smokery.

I would also like to thank Nick Stokes, Environmental Health Officer of North Shropshire District Council for his good advice and knowledge of the food industry.

I deeply appreciate all the hard work my wife Ann has put into this book: transcribing my words from tape-recorder onto paper for me and for taking so many of the photographs for this book – as well as for being there for me at all times.

I would also like to thank my publishers at Merlin Unwin Books, including Karen, Merlin, Jo and Gillian, for having the foresight to help me pass this knowledge on to the next generation.

Maynard Davies

Ann and Maynard, with a handful of the page proofs

The publishers also wish to thank Ann Purchase for her wonderful cooking at all our editorial meetings. Thanks also from Merlin Unwin Books to Anthony Bloor for his editorial help and to Phil Robinson of Dalziel Ltd for his expert reading and technical advice.

CONTENTS

A WORD OF CAUTION

This book records bacon curing as it was in my day. We produced our bacon and all the other meat products to the highest standards, using the best quality ingredients and on a relatively small scale. 'Good food for good people' was my mantra. Much has changed since then and the modern producer, if he is to sell his products on the open market, must comply with all kinds of new regulations unheard of when I was working.

For example, in my day we did not infuse our sausages with artificial preservatives. We used just the natural ones: the salts, the sugars, the spices (see my list of natural preservatives on page 132). Nowadays people expect to be able to keep their meat in fridges for days before eating and to a certain extent the modern food regulations are designed to allow this.

Quite apart from getting the right balance of natural preservatives in our meat products, I would always place great importance in the correct way of handling the meat, its preparation, and in its storage. I hope this comes across in my book. Good hygiene practices were as vitally important in my day, as they are now. But today there are different rules.

So if you intend to produce bacon, ham, haggis - whatever - and sell it today you must be aware of the prevailing laws and regulations, and use this book only as a reference point as to how traditional products were made. My best advice to you is to contact and befriend your local Food Standards officer. A good starting point is the following address:

Food Standards Agency (FSA),
Food Additives Branch, Novel Foods, Additives and Supplements Division,
Tel: 0207 276 8556   Fax: 0207 276 8514
website: www.food.gov.uk   email: foodstandards.gsi.gov.uk

The Meat Hygiene Service (MHS), an executive agency within the Food Standards Agency (FSA) is also a useful point of reference, as are the local Environmental Health Practitioners/Trading Standards Officers. And for guidance notes on Food Labelling Regulations and QUID (Quantitative Ingredient Declarations) see the following website:

www.food.gov.uk/foodindustry/guidancenotes/labelregsguidance/quidguid

INTRODUCTION

A Peep into the Past

I am writing this book of bacon curing recipes to pass on my knowledge to the next generation and to those who want to produce good food. Some of these recipes are of my own making; some are recipes I have been given during my 50 years in the food industry. I hope you find them interesting. I will describe the method of producing each recipe and I hope you enjoy the experience. Some of these recipes are traditional and old — so you will have to experiment and adjust them to suit the characteristics of your own meat and ingredients. Once you have tested a recipe, and if you are producing for re-sale, it is advisable to send a sample to the Public Analyst to make sure you comply with all the current regulations: legal requirements do change over the years.

Bacon curing is an ancient art, going back thousands of years, and it was practised by our ancestors to preserve pork for the winter. When the Romans came to England about 2,000 years ago, they brought some of the curing techniques we use today. The Romans traded with the Celts of northern Europe for hams, and the Celts were some of the first people known to have cured and smoked hams. Salt was readily available in northern Europe, and there was also an abundance of wood. The Romans, in turn, developed a taste for these northern hams, which they cured by dry salting. The salt the Celts used probably contained a certain amount of nitrate which, when it meets with bacteria forms *nitrite*, giving the ham a pink colour and helping with its preservation. In Rome ham-making flourished and the Romans produced very good hams. They had communal smoke houses, which have been uncovered by archaeologists. On one dig a large container was found with holes in the bottom and nobody could figure out why there were these strange holes in the bottom — it was one of the original curing vessels that had been unearthed.

Curing is a very old art on our own island; all the people that came to live here brought their own methods with them. The Vikings and the Saxons gave us a wealth of information. The Saxons used smoking as a way of preserving food; they had a primitive smoke house which is similar to the one that is constructed today, so curing in this country goes back a long way. The Romans showed us how to process salt and some of their salt workings are still in existence. The salt works at Maldon in Essex was reputedly opened by them and is still producing salt after two thousand years. When the Romans brought the technology to produce salt, the curing process progressed and that gave us the art to preserve food over the winter.

Throughout the centuries, we have perfected some delicious regional bacon. The pigs were reared and cured on the farm and some of the bacon was sold in the local markets. It was mainly the women in the household who cured the bacon; they also kept the recipes and passed them on to the next generation. We are lucky this happened as they left many of the old recipes that are still used today and there are some of these recipes in this book.

In the sixteen and seventeen hundreds a Master Curer went from farm to farm and cured bacon. His skills earned great respect because people depended on him to cure the bacon that would last for the winter; he would kill the pig one day, then rest the meat and cure it the following day. The killing was done in the dairies and that was the beginning of our curing industry.

Some of the hams from the Home Counties found their way to the London taverns. Ham and eggs was a very popular dish in the seventeenth century and there were inns in London which just sold ale and ham and eggs. So you can see how an industry blossomed but you needed great skill to practise it and the Curers were held in very high esteem in each area as they were an essential part of the community.

As the Industrial Revolution took place, the bacon industry progressed and there was a demand for food at a reasonable price. In 1770, the Harris Bacon Company of Calne in Wiltshire built a factory to produce bacon in large quantities to feed the population of London and the south. This is when bacon was produced in much larger quantities and when they started production they developed a bacon called the Wiltshire cure, which is still in existence today. In those days it was done by using the dry salting method and the wet curing method. The Harris Bacon Company had a Royal Warrant from King George to produce bacon.

As time progressed bacon was produced in Ireland. It was good quality bacon and the Irish were the first to improve the basic pig stock. They imported some very good Large White boars to improve the stock and so the bacon curing industry took off.

Now we come to the other side of the industry — **smoking bacon**. This has been practised for over 2,000 years. The Chinese and the Celts did it, but every civilisation had different methods. Here in the British Isles we used an oak chip (a hard wood). Smoking was done to preserve food, and the traditional way was to dig a large trench, fill it full of logs, put the meat or game in — hanging over the side on long poles — and cover it with branches. That is the earliest preservation method in England.

In this country, smoking bacon was done more in the south than in the north. I think one of the reasons for this was the

That is how the smoking industry began: primarily to preserve the bacon for longer and to keep it in good condition over the winter months.

Maynard Davies. 2009

The traditional smoke house, the likes of which have been in operation in England for centuries

fact that it was colder in the north and warmer in the south. The bacon kept better in the north and a stronger cure was used. Smoked bacon was mainly used in London and there was a particular way it was smoked. Most of the bacon went to London unsmoked; it had its own company that smoked the bacon. The bacon was smoked using oak chips and straw which gave the bacon a lovely colour. Juniper berries were incorporated in it which produced a lovely flavour and they called it the London smoke. The Welsh also smoked their bacon but they used peat. And the Irish also used peat.

PART ONE

TRADITIONAL BACON CURING

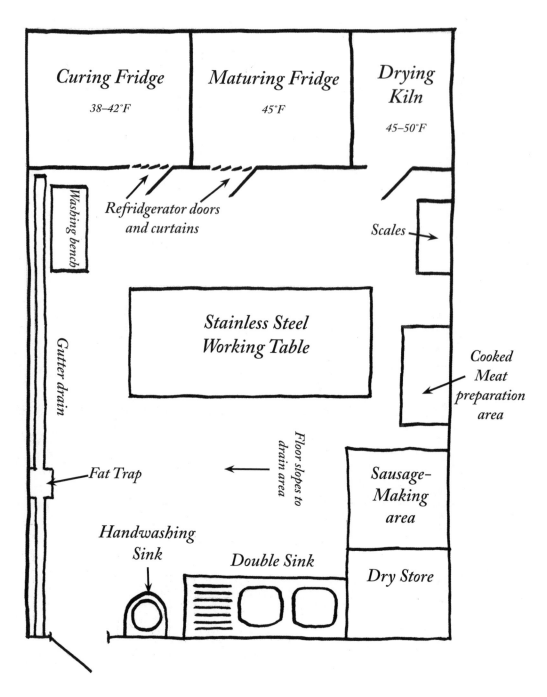

The meat curing house layout, in Maynard's day

CHAPTER ONE

THE CURING HOUSE

What makes a good bacon curer is not rocket science, but common sense. The whole idea is not only to make bacon but good bacon.

THE ESSENTIALS

One of the essential requirements is an area that is large enough to cope with your needs; I always feel it is better to make the curing area twice as large as you think you want to cure in, as it is expensive to expand it later. Your curing house should be easy to clean and easy to run. The walls, ceilings and floors should all be washable and cover, roughly speaking, an area of 15ft by 15ft — for a small producer of bacon, 20ft by 20ft would be ideal.

SLOPING FLOOR

I suggest the floor be slightly sloping to one end to a drain so that when you wash the area down the area dries quickly and there are no puddles to accumulate as this is a danger to the operatives. The drain needs to have a fat trap in it so the fat does not accumulate in the pipes and block them up. Also, to prevent the fat

and debris getting into the main drainage system, there should be an inner lining which can be taken out and emptied every day. The floor needs to be topped with granite chips. This gives it a hard surface which is essential because, without this, the salt will act as a corrosive agent and over a period of time the floor will disintegrate, but with a topping of granite fillings you will find that the floor will last for years.

THE HARDWARE

In the curing house you will need a large double sink (this is important) and a hand sink; on the walls you will need a fly killer. You will also need a large stainless steel bench which you will do all your curing on; this will be used for both wet curing and dry curing.

● Everything must be stainless steel as salt is very corrosive and it is a waste of time and money using any other material. I suggest you have waterproof electric plugs and they should be high enough on the walls but within reach to use and clean. I also suggest that the doors into the fridges be wide enough to take a hand pallet truck — this is essential — and the containers you cure in should have wheels so that you can keep everything on the move. A good-sized polythene container with wheels on is ideal for your curing house.

● The lights in the curing house must have watertight fittings to prevent condensation entering. I recommend fluorescent lighting rather than bulbs — if a tube ever breaks it will not scatter everywhere, and you will be able to replace the tube with no fear of glass getting in the production.

● You will need a toolbox to put all your tools in. You will need a full set of knives, a saw, a mallet and a curer's chisel. You will need a sharpening stone to keep your knives sharp; a steel glove and a steel apron for safety reasons; a stainless steel table and a small slicing machine; and a pair of scales as there is no guessing in this business.

● It is a good idea to keep a seasoning room for all your herbs and spices. You will need polythene containers with airtight lids and with the name of the ingredients on the lid and on the container. It is best to keep all the ingredients in a dry atmosphere.

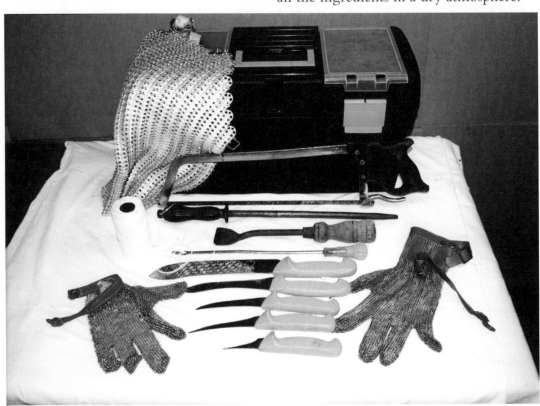

Tools of the Trade: the safety apron (top left), toolbox, saw, sharpening steel, curer's chisel, threading needle and string, steaking knives, and safety gloves.

My curing house: packing bench (on left), sausage manufacturing equipment and preparation tables.

● You will also need a brineometer for measuring the salinity of your brine (see the glossary), and it is best to have two — there is nothing more annoying than being halfway through a production when the brineometer breaks and there is no spare; then you would not be able to measure the strength of your curing solution. Always wash your brineometers after use and return them to their containers fully dried; they are an expensive piece of equipment so treat them with respect.

● You will need a thermometer in the curing house and also in the fridges and then you can keep check overall. In an ideal world, you could do with the curing house being air-conditioned and that would keep a steady temperature: it is an expensive investment but will recoup the cost many times. A hygrometer (which measures humidity in the room) is also invaluable.

Keep your own recipe book

You now have a curing house and are ready to start curing. One good idea is to have a recipe book of your own. If you alter a recipe make sure you write it down; also write down your mistakes so that when you want to look back you have a comprehensive list of mistakes and successes — do not leave things to guesswork.

Make sure your tools are always sharp, buy your own whetstone and keep them in good condition. Never put your tools away dirty. Always clean your own knives and never lend your tools to others because they will not sharpen their knives like you! Your knives will, over time, develop a unique and distinct 'edge'.

That is the basic set-up for a curing house, and here again I say the golden rule of the curing house is cleanliness. If you follow this golden rule, you will have an abundance of success.

MAYNARD'S TIP

No man can know too much about the tools he handles each day, nor be too big to refuse guidance from others who know!

Two essential tools for your curing house: the brineometer (above), and (below) the whetstone for knife sharpening

CHAPTER TWO

BASIC INGREDIENTS

SALTS

All natural salts have different densities, or levels of salinity, so any measurements are approximate and only the use of a brineometer will give an accurate reading.

A real friend to mankind, salt was first used in food preparation reputedly by the Chinese, who used it to salt fish. The ancient Egyptians used it to salt vegetables, so we have been using salt for many centuries.

Probably the oldest salt works in England are in Maldon in Essex, and Maldon is still today renowned for its excellent salt (produced by the Maldon Crystal Salt Company), as is Anglesey in North Wales, where salt is produced by the Anglesey Sea Salt Company.

It was only when the Romans arrived in this country that we learned how to process salt on a large scale. Before that we were very primitive: the early Celts would put salt water into a clay vessel on a fire, the water would evaporate and leave the small grains of salt. It would only be a

very small quantity. The Romans showed us how to produce more by using large shallow pans which trapped the grains, and that is how real salt production started.

The large-scale production of salt was the basis for producing excellent cured products in England — predominantly bacon.

The best salt I have found for curing was Bay salt. It comes from the coastline running from France to Spain and Portugal. In Napoleonic times, when the war with France was in full swing, apparently our famous York hams did not taste as good because our Curers could not acquire the salt from the Continent. I am not sure if that is true but it's worth considering.

Salts come in very different flavours (there is even a sweet salt) and in many different colours, including red, green, and blue — it just depends on which part of the world it has been mined in. The salt mined in northern Europe — Scandinavia, Germany, Poland, Austria — is a grey salt, and that has been the type we have used for centuries.

The first people to use salt in any quantity were the Celts in central Europe. They were the first ham-makers and they had all the ingredients to hand: salt was very near the surface, there was an abundance of timber, and there was also an abundance of wild boar. The Celts traded their hams with the Romans, and the Romans did like their pork — it was one of their famous dishes — so this was the start of the bacon trade.

Salt does enhance the flavour of bacon but it depends on which salt you use and which ingredients you add. Table salt often has additives such as iodine and anti-caking agents — these will affect the taste of your product, so I do not recommend the use of table salt in your cures or seasonings. A good knowledge of salt is essential as a lot of people do not understand the different flavours of salt. They are the poorer for that, because they are missing the finer points of curing bacon. If you understand the mechanism of salt, how to apply it and which grade to use for different products, that is the way to success. Knowledge opens many, many doors.

One of the most famous salts which England produced was known as Liverpool salt. It was exported all over the world in different grades, sizes and quantities. It went from the Liverpool docks but it came from the Nantwich area, which was an important salt area. We have a heritage in salt but our salt is quite different from continental salt, and the secret of curing is in the blending of the salt with other ingredients to achieve a fine flavour.

In writing this book, my mind has wandered back to the days when I served my apprenticeship. I remember the salt room. It was a room especially for salt, and we used a tremendous amount of salt as we produced every product imaginable. There was a different salt for every job and I remember there was a special way to stack the salt. They never laid the salt down in the bags. Instead, the bags were always propped upright, and being an inquisitive kind of young man with a lot to learn, I asked the old Curers why this was. One of them who had an abundance of knowledge told me: 'Maynard we stand the salt bags upright so there is a current of air that passes between them. This stops the salt from becoming damp.'

The salt bags were stacked on a stillage, which is a structure 6" off the ground with rods running across to support the bags of salt. This method keeps the salt dry, which is essential in the curing industry. So that was where all the salt bags were stacked and above them were the names of the different salts. Once a bag of salt was opened, it was put into an airtight container as damp salt alters the density of the curing mix. On the other side of the room we had the containers of open salt, ready for use, and that is how we did it.

I suggest that you also have a separate salt store in order to keep the salt dry and damp-free. It is the material you use the most and I think that if you keep it right, and give it a little bit of love, it will pay you back.

Bay Salt

This comes from the Mediterranean and has a distinctive flavour, which comes from tiny impurities in the salt. It is much used in good quality ham production and in other cooked meats: the large grains are particularly suitable for ham production and the reason for that is because the grains allow the fluid in the meat to drain easily.

If you use the finer salt on the other hand, you will find it will clog up and the meat fluids will not evaporate so quickly and you will have a soggy mess. By using the Bay salt you will have what is called a 'free flow.'

For sausages, the large grains must be milled down to a finer consistency and this can be done by putting them in a grinder. The salt can then be put into cooked meats, sausages and spice brines to give you a distinctive flavour. I recommend Bay salt as an excellent product in our industry.

Roman Salt

This salt comes from Italy, Sardinia, and along the Mediterranean coast. The salt has a small amount of nitrate in it and it is used in the production of Parma ham. When applied to the ham it gives it a pinkish colour. It is also used in the production of salami. The larger grain of Roman salt is used for the Parma ham and the smaller grain for the salami, pancetta and sausages.

Dairy Salt

This is a strong salt and can be bought in bars or in bags. It is used, as its name implies, in the dairy industry for butter and cheese. It is also used in the curing industry for different products: it can be used in sausage manufacturing and in brines. It has a distinctive flavour all of its own.

Lump Salt

This is a general-purpose salt and is used mainly for putting on the beds and stacking bacon on top. It is an economical salt for general use in the industry.

Rock Salt

This is a salt that is mined. It is the impurities in it that give it its distinctive flavour. It is used as a high-class table salt. It comes from many parts of the world — Germany, Austria, Poland — and also from England. It is an unadulterated salt and you must decide which product you think will be enhanced by its use.

Sea Salt

This is a natural product from the Mediterranean and the Black Sea. There are many flavours of sea salt and you have to decide which one you think suits your product. At one time I had five or six different kinds of sea salt, each of which suited a different product. We even had a sea salt from Malta that was a very strong salt and it did very well in the salamis.

But do your own tests and make sure you mark the containers with the different strengths. Sea salt should not be confused with Bay salt.

Vacuum Salt

This is pumped from the ground as brine; the water is then removed and the salt is left. Normally an anti-caking ingredient is added, such as magnesium, to aid the flow. I found that this salt (and this is only my opinion) was not a good salt: it lacked flavour and sometimes when we used it the brine would become very cloudy. I never achieved good results, but that could just be me, so do not let me stop you trying.

Saltpetre

Saltpetre is a preserving and colouring agent which, as nitrite, gives meat a pinkish colour. It is a preservative that inhibits the growth of bacteria, and must be handled with great respect. It must be kept on its own in a locked cupboard, clearly marked on the lid and side of the container. The chemical name for this is potassium nitrate (food additive E252) and when used in a product it must be tested by the local analyst for strict adherence to Government guidelines. An accurate pair of scales is essential and, when using saltpetre, always analyse your brine to check you are not using more than the regulation levels of nitrate. Saltpetre is very much stronger than most other salts. It can be bought from butchers' suppliers or salt suppliers.

SUGARS

There are various sweeteners used in curing, including sugars, honey, and treacle. The following are the ones I favour using:

Light Muscovado Sugar

This is a delight to use. It is a natural cane sugar and has a distinctive flavour, and gives good results. The best Muscovado, in my view, is the one from Mauritius.

From left: Demerera, dark Muscovado, light Muscovado, molasses sugar. Your choice will greatly influence the flavour and colour of your cure

Dark Muscovado Sugar

This is another fine sugar but be careful which product you use it with as it will darken the product and you may not want it to do that. However, for flavour it is King; a very rich flavour which will enhance your product. It is an excellent sugar to use for sweet brines.

Molasses Sugar

This is an excellent sugar and particularly good in spice brines and sweet brines. It is difficult to use in other recipes due to the quantity of molasses in it. I used to think it was also a good preservative. Highly recommended.

Demerara Sugar

A delightful sugar with a good flavour, used in most ham production and in cooked meats because its light colour does not discolour the product. Many uses in the trade — recommended.

Fine Demerara Sugar

Excellent for production, as above, and I find this is particularly good for the brines for smoking. An unrefined Demerara gives you excellent results.

White Granulated Sugar

I was never very pleased with this sugar and I only ever used it in an emergency. In my opinion, you should always try to use another sugar. I never had any good results with this sugar. It is a bit too refined and lacks character.

Honey

This was the original sweetener for bacon. It was also used in sausages and acted as a preservative. I have always used the clear light honey, but there are many types of honey (e.g. acacia honey) and these may be worth trying.

Treacle

Black treacle gives the bacon a very rich, distinctive flavour. It is used in the famous Bradenham ham, a black ham reputedly developed in Wiltshire by Lord Bradenham in 1781. It was later produced for the wider market by the Harris Bacon Company in Colne.

Golden Syrup

An ingredient used in bacon curing to give a very sweet flavour.

Maple Syrup

Used extensively in the United States and Canada in bacon and sausage production. When the first settlers went to North America, Maple syrup was an easy source of sugar which was used to a large extent instead of honey.

SPICES, HERBS AND BINDERS

It is important to understand spices, herbs and binders as they play an essential part in your business. To understand them is to unlock all the knowledge you will need to produce different flavours, as one spice is very different from another, each with its own individual flavour. The idea with spices is to marry them together so they complement one another. It is important that you understand the products you use, so I hope this chapter will help you and enable you to blend your own herbs and spices and put your own name on your mixes; this will play a tremendous part in your business.

You will need a good recipe book, a place to prepare the ingredients, and an accurate pair of scales. There should be no guesswork in the adding of spices and herbs as this road leads to nowhere. This is the hub of your business and you must really take to heart the products you use. Nothing is more disappointing to your

The pestle and mortar (left), has been replaced by the modern grinder (right) for blending herbs and spices

customers if this week's product does not taste the same as last week's. She depends on you to produce the product she enjoyed the week before, so it is up to you to keep correct records and follow your recipes carefully.

Make sure everything in the seasoning room is correctly marked and do not order more than you need because the spices and herbs will become stale. Always keep them in good airtight containers. A grinder is a good thing to have, as this will crush your herbs and spices down to a fine consistency. I also found a coffee grinder worked well. I bought a commercial coffee grinder which was an excellent machine. It ground the spices to medium, fine, very fine or how I wanted them, so in the long term it is a worthwhile investment when you have a lot of herbs and spices to mix.

For most other mixing jobs I recommend using a bowl chopper: this is one of the main machines in your factory and it is ideal for mixing larger quantities. But for herbs and spices, I think the coffee grinder is better. Always wipe out the bowl after use but do not let water into the mechanism of the grinder.

You can buy herbs and spices already ground and packed in smart containers, but I always felt I wanted to know what I was making and I also felt that the customer deserved the best. I think if you mix your own herbs and spices you have your own distinctive product and you will get well paid for that. I suggest you adopt the slogan: 'We make all we sell and we sell all we make.' That was my motto and if you follow that you will be taking the road to success.

Pepper

Pepper is a good bedfellow to salt. 3oz pepper to 1lb of salt will give you a good balance for your seasonings. I used white pepper in sausage manufacture as I found the black pepper gave the sausage a grey hue so I've always used white pepper and in most cooked meats as well. Black pepper is better in salami, and certain cooked meats such as luncheon sausage and black pudding. I always marked the black and white pepper clearly and never ground too much at one time. Just grind enough for the week's production. I found if I ground too much the aroma faded very quickly. When grinding peppers put a small amount of rice flour in the grinder and you will find the rice flour will trap the aromas and the oils and keep the strength in.

Coriander

In my opinion coriander seed is the prince of spices. The Romans originally brought coriander to this country. It is an aromatic spice and is used with great abundance in the food industry: you can use it in pork sausages, salamis, the curing of hams and bacon. You can also use it in spice brines. It is a spice for general use which gives the product a gentle, sweet-smelling taste. It needs to be used to a ratio of 1oz to 1lb of salt — finely ground in your sausage, this will give you good results. Coriander mixes well with nutmeg and mace, to a ratio of one, one and one. It is a superior spice and I used it a lot.

When using coriander in sausage seasonings it is a good idea to put the coriander with mace and nutmeg and to add a teaspoonful of rice flour. You will find this will blend the spices better — they seem to marry one another.

Ginger

An excellent spice in tomato sausage and it blends well with mace and nutmeg. The ratio should be equal parts: one, one and one. It is also added to polonies, savouries, luncheon meats and beef sausage. I have also used ginger in bacon curing which gives the bacon a very pleasant taste.

I recommend grinding the ginger with a small amount of rice flour prior to adding the other spices. Root ginger is a

MAYNARD'S TIP

When roasting a leg of pork grind some caraway seeds, mix them with apple juice and, half an hour before the pork is ready, pour the mixture over the joint. Then complete the roasting. This will give it a lovely sweet flavour.

hard spice to grind, so you need to grind it first to break it down before adding the other spices.

Nutmeg

A good spice for sausage-making and for cooked meats. It has a lovely sweet smell, a slightly bitter taste, and is an excellent bedfellow to mace. In fact, they come from the same tree: nutmeg from the seed, mace from the lacy covering of the seed. Mix with mace on a one to one basis.

Cayenne Pepper

A very strong pepper and you have to be careful how you distribute it. You will only need ¼oz to 1lb of salt. Ideal in a beef sausage recipe, and for dusting hams (mixed with white pepper in a 2 to 1 ratio — two ounces of white pepper to one of cayenne). It is also a very good spice with continental meats, but remember it is a very strong pepper so do use it sparingly.

Many people use cayenne pepper mixed with olive oil and painted on the hams as it stops mould and also discourages the flies.

Caraway Seeds

Used in continental sausages and to a degree in fresh sausages. It can be incorporated in the curing of hams and bacon. It gives a distinctive flavour and is best ground very finely. Use about 1oz to 1lb of salt when putting into a seasoning. The difficulty with caraway seed is to find a suitable spice to match it with. It is usually better to have a rather bland spice with it (e.g. mace or nutmeg).

Pimento/Allspice/Jamaican Pepper

Pimento, commonly known as allspice, or in some very old recipes, Jamaican pepper, is said to combine the flavours of cinnamon, nutmeg and cloves. It can be used in fresh sausages, beef burgers, frankfurters, liver sausage, and most cooked meats. It is used extensively on the Continent and in North America, and it is also used to a degree in England in fresh sausages and in luncheon sausage. It is a 'peppery' spice.

Juniper Berries

Juniper berries have been used in food production for centuries and, of course, gin is also flavoured with them. Juniper is put into brine for tongue, pork and spiced beef. It was traditionally used in the 'barrel' pork for the Navy, as it gave the meat a nice flavour. Juniper berries are also good to use in the smoke house. Mix them with coriander seeds and, when the smoke oven is going well, throw them on the burning wood to give the bacon a lovely aroma.

Cinnamon

One of our oldest spices, you can use cinnamon in fresh sausages and continental sausages and it is much used in the bacon industry, in particular for pie seasoning. It can also be added to meat pastes. I would use ¼oz of cinnamon to 1lb of finely ground salt. It is a warm, sweet seasoning and thoroughly recommended.

Cloves

Cloves are used in ham decoration. Before cooking the ham, the top is scored to make a diamond pattern. Whole cloves are then inserted into the diamonds. You can also add cloves to brawn, black pudding and meat paste. For the latter, make sure you grind it finely. Test it out for strength yourself and choose the right balance for the mix.

Bay Leaves

I used to have two bay trees in my garden so I used the fresh leaves a lot. I used them with boiled hams and with roast hams. I would grind them fresh and put them in the spice brines and they definitely produced a good flavour. I recommend them.

Celery Seeds

These can be used in sausages, continental sausages, liver sausage and blood sausage (not to be confused with black pudding). I always use celery seeds with caution, as they can have an overpowering taste. I found ¼oz to 1lb of finely ground salt a good mix, but here again I would choose the product carefully because the strong flavour of celery seeds can overwhelm some meats. I used them only occasionally. On the Continent they are used more widely.

Paprika

I can recommend paprika for pork sausage, beef sausage, some cooked meats and in bacon curing — the results are always good. It is an excellent spice in tomato sausage. To achieve good results, add ¼oz of paprika to 1lb of finely ground salt of your choice.

Mustard

Mustard can be used in beef and pork sausages. The seeds are black and white and they are ground together to make a flour. Only grind small quantities as the aroma fades quickly. I recommend a mix of ¼oz of mustard to 1lb of finely ground salt.

Garlic

A very strong bulb which must be handled carefully. You must make sure the quantity of garlic you use in your product is low. It is used in hamburgers, liver sausages, and on some salted beef and hams. I suggest you never grind the garlic in your grinder as the smell lingers on and can be passed to the next spice used in the grinder. So I always use a separate garlic press.

Onion

Onions can be used fresh, kibbled (dried), or in powder form. I preferred to use fresh onion as it gave a better flavour, making sure it was finely cut with a designated knife, for, as with garlic, the flavour can transfer easily to other products. Onions can go in black pudding, salamis, hamburgers and luncheon meats.

MAYNARD'S TIP

Kibbled onion should be reconstituted with water before use; powdered onion should be used very sparingly.

Herbs

You can grow sage, thyme, marjoram, basil, parsley, pennyroyal, mint and bay leaves in a herb garden. You have to be careful when you harvest your herbs that you do not bruise them, and you must dry them carefully and store them in an airtight container. Grind them to your recipe when you need them. Using fresh herbs is nice to do in the appropriate season and it will enable you to get a good price for your sausage. There are a lot more herbs than the ones above which you can introduce to your seasonings, but again it is up

to your taste. I would recommend ¼oz of dried herbs to 1lb of finely ground salt of your choice. The seasoning room is the hub of your factory so you must ensure that the weighing is done by a responsible person. It should not be a haphazard exercise. The flavourings should always complement the meat, not overpower it or mask it. Remember: with seasonings you can always add but you can't take away. I always had one rule: if I could not eat it then nobody should.

Binders

Binders are added to sausage meat to absorb the fats and juices which are released from the meat when cooking takes place. The juices contain the flavour and you do not want to lose these in the cooking process. The original binder for sausages, pork pies and cooked meats was boiled rice. It was in the 1914 war, when rice became unobtainable, that bread and rusk were introduced as alternative binders.

Patna rice is one of the best binders for it stops shrinkage in the meat. It is boiled and then mixed in a ratio of 3 to 1 (three of water, one of rice). This extends the shelf life of your product. Make sure your binder is always stored in a dry place. As a general rule the binder is added to the mix and then blended with the sausage meat. A mix of rice and rusk also makes an excellent binder.

Rusk

Rusk is commonly made from a yeastless flour, mixed with water and baked. It can be bought in many grades — large, medium and pinhead — and it helps to absorb water and meat juices. Place the rusk in a bowl and pour cool, sterilised water over it. Leave for half an hour before adding it to the meat. When we added rusk we always used the pinhead rusk as it blended in well. If we used a large grain rusk we found it took too long to grind to size and made the machine hot. Warmth is not a good thing when sausage making because the fat content becomes too sticky.

Flour

This is another good binder. We used to use flour and cornflour in equal quantities. By adding that to the luncheon sausage and the black puddings and liver sausage it gave a tight mix and enhanced the slicing qualities, so I do recommend flour and a certain amount of cornflour with it and in some cases a small amount of rusk can be added too.

Farina

A good stiffener, to be used sparingly. Farina, in British curing, refers to potato flour. It has a pleasing flavour and has good binding qualities. About 1lb of farina to one pint of water is the right mix. Too much farina and you'll have a spongy sausage. Do not use in raw sausage meat as it will make the mix too stiff.

Cornflour

A starch product which stops the meat discolouring and is a good absorber of water. The best way to use cornflour is to mix it as follows: one part of cornflour to two parts of ordinary wheat flour and one part of small-grain rusk (and one pint of water if necessary). Cornflour is good for luncheon meat and liver sausage.

Soya Flour

Soya flour is made from roasted soy beans ground into a flour. It is a source of high-quality protein and has excellent binding qualities. It is good in fresh sausage and cooked sausage. A lot of people put it in hamburgers and pork burgers for it marries very well with meat and improves the slicing qualities. It can be used as 1% of a dry mix. Do not add more than 1% or it will mask the true taste of the meat.

Milk Powder

This is an excellent dry product to put in fresh sausage. It gives a nice creamy texture and is a natural preservative. The proportion you will need for a dry mix is about 1% milk powder — not more than 1% as this will make it too creamy. There is a distinct advantage in using milk powder in sausage production as it produces a nice

MAYNARD'S TIP

Store milk powder in a very dry place. Keep it in a polythene bag, tie the top and then store in an airtight container. Do not order more than you need. It is a difficult product to keep because it takes moisture from the air and goes into a block.

smooth product, absorbs the water and meat juices, gives the sausage a pleasant colour, and enhances the product.

Polyphosphates

Polyphosphates are chemicals which are used commercially to retain water in meat products, thus making them heavier; and the customer is paying for that water. They are also chemical flavour-enhancers which are used in meat production to combine water, meat juices and fat. They are put in most 'bought-in' sausage seasonings, pie seasonings, and commercial bacon curing mixtures, which is why it is much better to use natural ingredients and make your own. I am not a great believer in using chemicals like this. Put an ounce of polyphosphate in a glass of water and watch it turn to jelly — enough said.

Monosodium Glutamate (MSG)

This chemical enhances flavour and the Chinese are fond of using it in their cooking. A form of it can be extracted from seaweed and has been used for centuries in Asian cooking, but MSG was not isolated and scientifically understood until 1907.

The disadvantage with it is that it not only brings out the good flavour but also the bad. I do not recommend using either polyphosphates or monosodium glutamate; I much prefer using pure ingredients.

Salts (above) come in a surprising variety, each one distinctive in terms of colour, size, strength and taste.

CHAPTER THREE

THE PIG

Bacon curing starts with the pig. Your pig must be of good quality and it is better to pay a higher price for a good quality pig than to cut corners and buy a cheaper, poor quality pig. I always think the best pigs for curing are the Large White, the Gloucester Old Spot, and the Large Black. The British Saddleback is also a good pig for curing as it gives you a good layer of fat, and in bacon curing it is the fat that holds the flavour.

A group of Large Whites (13 weeks old) in conditions which allow exercise and fresh air. Pigs are sociable creatures and always do best when reared in groups

Which pig and what size?

The best pigs for curing are in the region of 200lb live weight. You will lose about 50lb from slaughter to curing. A gilt (female) pig is better to use because if you use a boar (male) there is a likelihood of developing 'boar taint' on your product.

Pigs should be starved 24 hours prior to be taken to slaughter and given plenty of water. Put some sugar in the water and this will encourage them to drink. If you take pigs to slaughter on a full stomach the meat tends to be dark, whereas starving them for 24 hours ensures the meat is a nice light colour. The best thing to do is to take two or three pigs at a time to the slaughterhouse as pigs are very sociable creatures and if they are on their own they tend to become excited. If they get excited there is adrenaline in the flesh and the cure does not take as well. If it is very hot weather, it is advisable to spray them with a hosepipe to cool them down. If your pig is put in pens with other pigs it is a good idea to spray all the pigs so they all smell the same — this stops them fighting. So take a bottle of vinegar with you, or a cheap air freshener, and also colour mark your own pigs so they are not mixed up.

Once slaughtered, the pigs should be left at the slaughterhouse for 24 hours so the body heat dissipates. If you do not do this you will run the risk of the pigs developing bone taint and if this happens the meat will be wasted and you will not be able to use the pig. Also, ask the slaughterman to cut each pig into two sides.

The ideal carcass will have a light head, a light shoulder, a long straight back, and a full ham (back leg) right down to the hock. When the pig is hanging on the hook you will need to look for a u-shape rather than a v-shape between the legs; this will indicate a good formation of the pig. The weight of the pig, ideally about 150-200lb, is important and the flesh should be a pale pink, the fat white, and the rind a light colour. These indicate quality in the pig and that will be echoed in the quality of the bacon you will produce.

I will now give you the different weights of pigs as all businesses are quite different and the weight of pig that suits one may not suit another.

As a general rule, the small- and medium-sized pigs tend to be sold by butchers as joints of pork, whereas the larger pigs go into manufacturing: the legs are cured, the shoulders made into sausages, and so on. Basically, you can't make good sausages from small, young pigs: there is too much lean meat and not enough 'binding' meat and fat, and the sausage skins would burst because the young meat contains too much moisture.

The smallest pigs will be between 60-80lb. These are known as 'porkers' and they can be cut into joints. There is very little waste on them. However, in my opinion, these pigs — although very lean and tender — do not have a lot of flavour.

The next size category is 80-100lb. These are known as 'porkets.' There is no waste on these and trimming is minimal. The next on the list is the 'cutter' which weighs in at about 100–140lb. The cutter is used for fresh joints and manufacturing.

The next size up is the 'bacon' pig, which runs from 140–160lb. They can be utilised in many different ways: the middles and hams can be used for curing and the streakys for manufacturing, but here again it's horses for courses — you choose the pig that suits your requirements.

The next category is the 'overweights' — pigs that are 160–225lb. These are female pigs, over a year old, which have never bred. In my opinion it is only once a pig weighs 160lb that it becomes ideal for sausage-making. The meat has sufficient fat to hold a good sausage together.

Then we have mature sows — female pigs that have had piglets. These are also useful for manufacturing products as the meat has better binding qualities.

> **MAYNARD'S TIP**
>
> It is a good idea not to put young pork into manufacturing as the meat has a high percentage of water and when used in sausages it causes bursting of the skins. Dark coloured pork denotes bad bleeding in the slaughterhouse and makes for an unpleasant taste. Do not hesitate to reject this pork, as it will not do your business any good.

The boar pig, which is used for breeding in the pig industry, is not such a good proposition as a source of meat. I have never been a lover of boars. I have always thought they were more trouble than they were worth, with 'boar taint' being a problem with their meat. However, the manufacturing industry buys them and puts about ten percent into sausage manufacture.

A Large White sow with cross-bred Tamworth piglets. A contented pig, raised in good conditions, is a much better prospect for the bacon curer

Checking some hanging halves for quality

The Cutting

The way in which the pig is cut is a personal decision, so I am not going to lay down any fixed rules and regulations for this as everyone cuts for their own business. However, if you do not get involved with the slaughtering you may find it useful to know the names of different cuts, so if you buy from a wholesaler or from the abattoir you will know what to ask for.

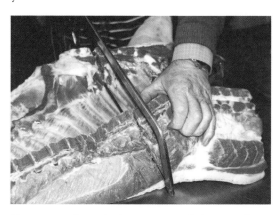

Everyone has their own special way of cutting and preparing pork. With experience you will work out which cuts work best for the particular cures you seek to achieve

Cuts of pork (uncured)

Carcass — the whole pig with head, shoulders, middles and legs.

Side — a carcass which has been divided right down the spine to make two symmetrical halves, or sides: each side with a shoulder, middle, leg, and half a head.

Middle — this may be split into two halves: the belly and the back.

Cuts of bacon (cured)

Spencer — consists of the shoulder and middle.

Three quarters — consists of ham and middle.

Side of bacon — leg, middle and shoulder (but no head)

Long cut — includes the collar and the back; it is half the shoulder and the back.

Middle — this is the gammon and the shoulder removed which just leaves the whole of the middle section.

Back — this is the loin minus the ribs.

Streaky — the belly and the flank, which has been taken off the middle.

Fore-end — this is the shoulder, or front end of the side of bacon.

Gammon — hind leg of the pig. It is known as a 'gammon' if it is cured as part of a 'side' – see above. However, if it is cured on its own, as a hind leg only, it is known as a 'ham'.

Those are the basic cuts. Sometimes the fore-end is cut into a picnic ham — this is the shoulder (with the blade removed) and shaped like a ham (see photo on page 30). I used to sell a tremendous amount of these at Christmas because they were econom-  ical to buy. We used to store them from the end of September and they were a very good buy for Christmas. I suggest you think about making them; they are very good value for the customer.

There are some cuts of pork you may need to buy in to keep your stocks in balance —a middle, a streaky, a back, a shoulder, and a short cut leg. These cuts you may have to buy in from the wholesaler to keep the stocks right.

The keynote to successful food manufac- turing is consistency. To achieve this you must have a complete knowledge of the curing process and the raw materials. Inconsistency has a dire effect on your customers as they always expect the product to be the same as the one they bought last week!

Always buy your carcass on grounds of quality and not on price. If you do this, then there is no reason why you should not produce a first-class product, one you should be proud to produce and sell.

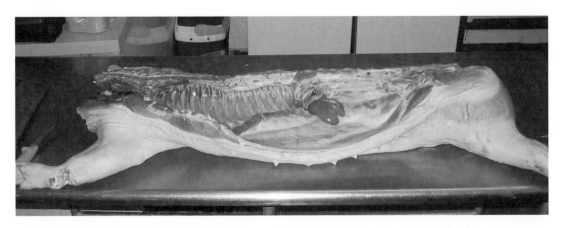

The side of pork as it usually arrives from the abattoir showing the kidneys in place. A side like this should always be hung from the back leg

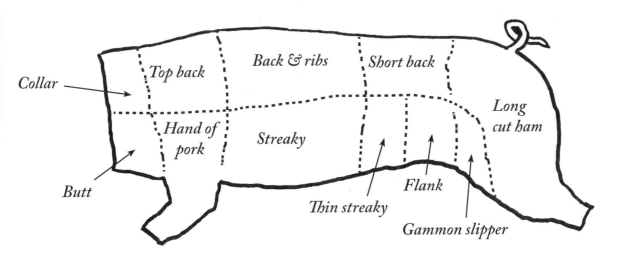

The Butcher's cuts of bacon vary from region to region. The cuts marked here are more detailed than would normally be of concern to the Curer

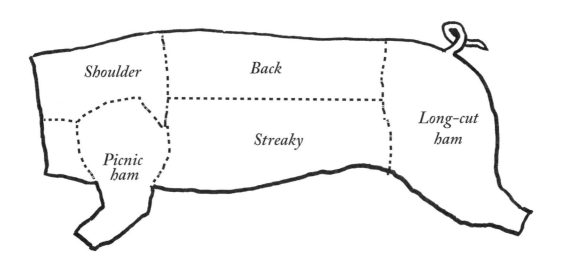

The Curer's cuts of bacon: (left to right) the shoulder and picnic ham, back and streaky, and the long-cut ham

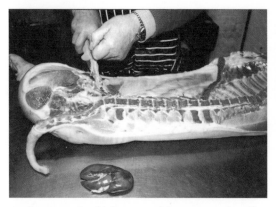

1. Cut out the kidney and locate and pull aside the main artery into the leg

2. Removing the leaf or flare fat

3. Cutting the leg off

4. The leg has been removed: now for the shoulder

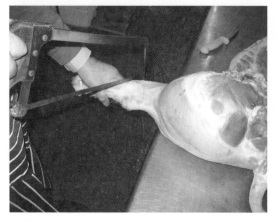

5. Removing the trotter

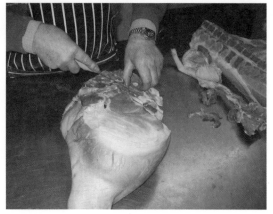

6. This trimmed leg is now in readiness for curing

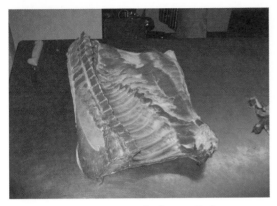

1. Middle section ready for splitting. Alternatively, it can be cured as one piece

2. First saw through the ribs

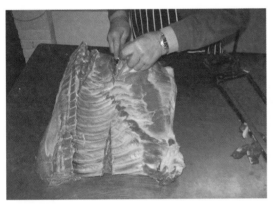

3. Now cut through the middle

4. Two halves: the back and the belly

5. Removing the ribs: the first cuts

6. Cut as close to the ribs as you can

7. Cut the 'rib sheet' away (it can be used for ribs)

8. Ribs, back and belly, now ready for curing

PREPARING THE SHOULDER

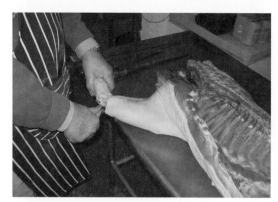

1. First remove the trotter

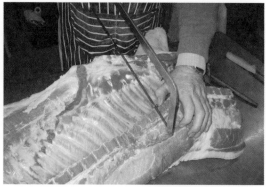

2. Saw through spine to separate the shoulder section from the middle

3. Now cut along the ribs

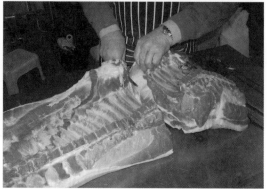

4. All the way through until the sections are divided

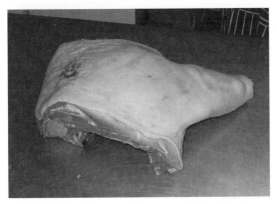

1. The shoulder section ready for further refinement

2. Cutting out the neckbone and ribs

3. First cut off the half-shoulder

4. The picnic ham (left) and the straight side (right)

5. Now remove the blade

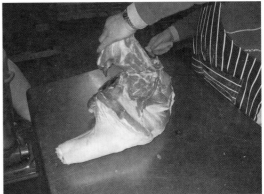

6. Cutting carefully around the shoulder blade

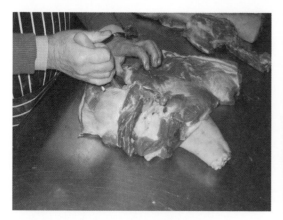

1. Boning out the shoulder blade

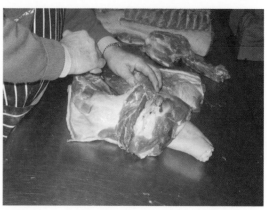

2. Working around the bone – always cutting away from fingers

3. Carefully releasing the bone

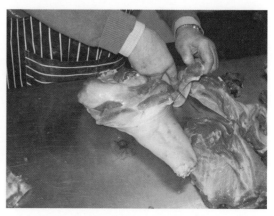

4. Trimming the picnic ham

5. Picnic ham in progress

6. The picnic ham and all the trimmings

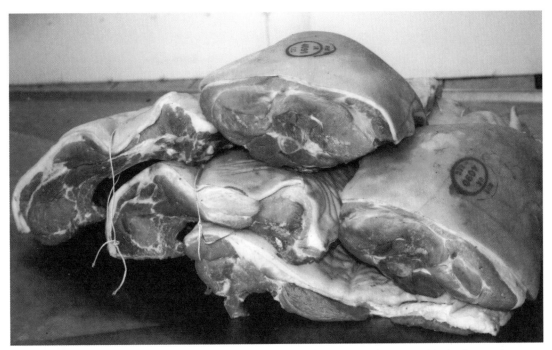

Hanging strings in place and the ham (top centre) and middles are ready for curing.
Below: *Maynard with a good batch of cured middles ready for the smoke house. At this point you should remove the middles from the fridge and allow them to come to room temperature, getting rid of any surface moisture before the smoking process begins*

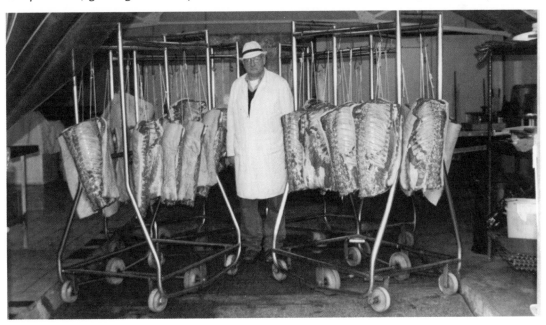

Tools for the bacon curer (from top): ham gouge, stainless steel skewer, curer's chisel

Two sharpening steels

MAYNARD'S TIP

When removing the femur bone, I suggest you use a curer's chisel. It does a better job and keeps the shape of the meat.

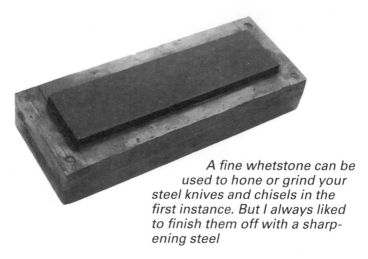

A fine whetstone can be used to hone or grind your steel knives and chisels in the first instance. But I always liked to finish them off with a sharpening steel

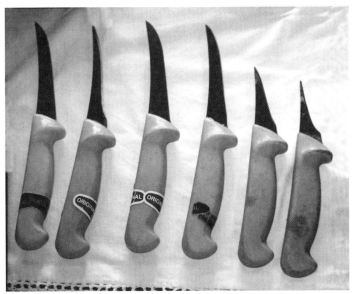

A selection of boning knives: some have stiff blades, others have a bit of flexibility

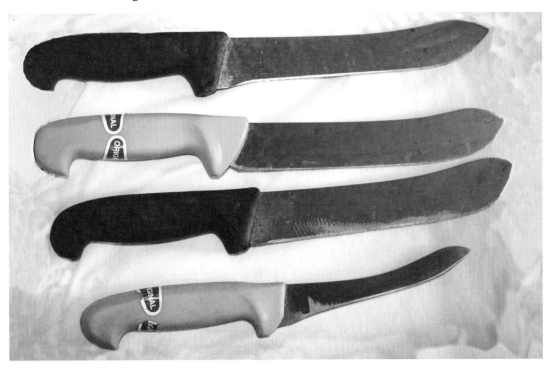

A selection of steaking knives, the smallest one is a three-quarter knife

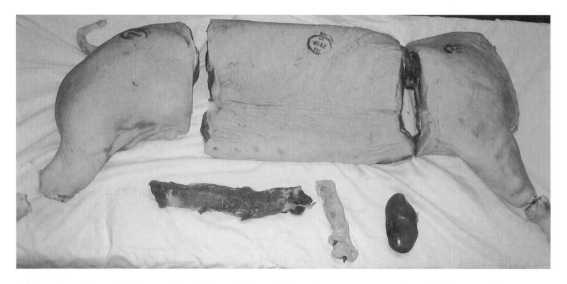

Side of pork ready for curing. From left to right: the gammon, the middle and the shoulder. Below: the tenderloin, flare fat and kidney

CHAPTER FOUR

HOW TO CURE BACON

There are many ways of curing bacon but I will start with the traditional way, and then I shall tell you about the modern way of producing bacon.

The most important rule in curing is to keep your curing area and your utensils spotlessly clean and also to wear protective clothing. Cleanliness is the golden rule of curing.

The ideal temperature for curing is 35-40°F. If it is too cold the process of curing stops and if it is too warm spoilage can develop (in other words, the wrong sort of mould will appear on the bacon).

Cleansing Pickle

The first thing to do when you start curing is to make a **cleansing pickle**. You will do this after the pig has been cut and is ready for curing. The function of the cleansing pickle is to clean the pork of all debris including bits of bone and blood.

The cleansing pickle is made using 10 gallons of water, 10-15lb of rock salt, and 4oz of saltpetre. Put the water and salt into a clean container and mix well — this is the basis of the cleansing pickle. Now put the saltpetre in a saucepan with a small amount of water, heat until dissolved, and leave to cool. When cold add to the cleansing pickle and stir well. Leave pickle for 24 hours in the fridge — it is then ready for use.

Put your pieces of pork in this solution for the following times:

Hams and Shoulders — 60 minutes
Middles — 30 minutes
Streakys — 10 minutes

MAYNARD'S TIP

It is always better to leave your pork in the fridge for 24 hours prior to curing, so that the meat is the same temperature as the brine.

35

As you put the pork in the cleansing pickle it will turn pink and all the rubbish will fall to the bottom. This pickle cleans the pork of all debris and the pork is now ready to cure — always throw a cleansing pickle away after use and do not re-use.

Black hams maturing at Duke's Hill

Curing Ingredients

Now we must consider the curing ingredients. The main ingredient for curing is salt. For wet pickles, such as the cleansing pickle I have just described, the best salt is rock salt. Rock salt has large granules and dissolves slowly, giving the ham a lovely taste. That is why it is used in the production of York hams. There are alternative salts such as sea salt, which gives a gentle flavour, and Bay salt, which is not so harsh.

The other key ingredient is sugar. For sweet cure pickles I would choose molasses sugar as this has not been refined and has retained all of its flavour. For a dry salted bacon, I like to use the light Muscavado sugar, or sometimes even a Demerara sugar. The choice of these different sugars depends on individual taste. Herbs — like coriander, mace, nutmeg and ginger — also play an important part in the curing process. Wine, honey, golden syrup, treacle, cider and maple syrup are just a few of the many other ingredients one can use.

MAYNARD'S TIP

Always keep the same saucepan for the saltpetre, never use it for other ingredients as cross-contamination may occur.

Wet Curing

Wet curing involves the use of a brine, such as the sweet pickle described on page 39. You will need a good container, twice as big as the volume of the ingredients. When you put middles into the container, you will find the brine will rise up, so the container should never be more than half-full and always be made of stainless steel or a strong plastic material. Thoroughly clean containers before use and boil all the water you are going to use for your brine and then leave it to go cold.

Boil your sugar with water and then leave that to cool; add your salt, stirring all the time to dissolve the salt thoroughly. It may take two of you to do this, one tipping the salt and the other stirring. The salt-petre should have previously been boiled until dissolved and left to cool. Then this is added to the brine.

When you have made the brine it is a good idea to leave it for 24 hours to cool down and to ensure all the ingredients are well mixed. If you use the natural sugars these may leave a dark brown froth. To clear the froth take a sieve and remove it, then stir well.

Keep your brine in a cool environment, about 35-40°F. This is the ideal range of temperature: if it gets higher than that you may find the bacon will spoil; if it goes below that you will find the curing process will stop. Ideally, it should be near the top end of the range; this will make a gentle cure. Brine curing is a lot quicker than dry salting as the penetration is a lot faster.

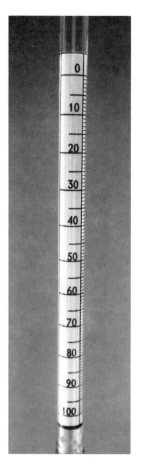

The brineometer (above) and a detail showing the scale (right). This useful tool indicates the strength of the saline solution by floating in it, rather like a fisherman's float. The line of the surface of the brine indicates the salinity of the brine

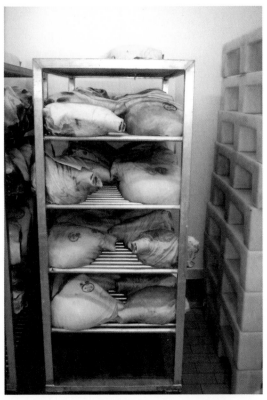

Hams on a rack in the maturing fridge where the temperature should be about 40-42°F

After 24 hours place all the meat in the container, making sure the meat is below the brine level and fully covered. One method I use is to place a slightly smaller lid on top of the meat with some weights on it. This will keep the meat below the brine line. If any of the meat is not covered then you will get uncured patches of meat, so it is important that all the meat is below the surface of the brine.

It is also important to overhaul the meat — this means to move the meat around in the brine every day or so during the process.

The next step is to choose how long you want the meat to be in the brine for. This depends on how strong you want the taste to be, and to get this right comes with experience. Usually it is about a week to ten days depending on the taste you are aiming for.

When the time comes for you to take the bacon out of the brine, take it out, wash with cold water and dry it off. This step is important: it *must* be washed thoroughly with cold water. If you leave any salt or brine on the meat, you will find the salt will attract moisture, and mould will grow on your bacon. Now place the bacon on a rack in a cool fridge (see picture on this page).

The bacon will now take about 12 days to what we call 'equalise' or 'mature' — this means that the salt and saltpetre will work their way through the meat.

After this the bacons must be taken from the fridge and hung up in a dry, airy place at a temperature of somewhere around 50-60°F. The bacon should be left here for at least another week during which time the maturing process will continue to work and the meat will gather flavour. It will then be ready to eat.

Obviously, the shoulders and the hams will take longer in the pickle than the middles — the bigger the ham the longer it takes in pickle. As I have said, it all comes with experience. Maturity is the key word in curing. When the bacon comes out of a sweet pickle it is *not* cured, so, as I have stressed, you must leave it in the fridge and then in an airy room to finish the maturing process.

After about a week you will find a

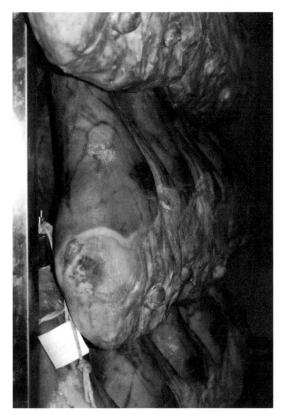

Rows of hams maturing. Note the white bloom of the 'good' mould just beginning to show

white mould begins to appear — this is a **good** sign that everything is progressing well.

If you do not have any refrigeration, the only time you can cure is in the winter months when the temperature is low. If you invest in a small fridge — ideally, a walk-in one — you will be able to cure all the year round and keep the temperature exactly as you want it, around 35-40°F for the brine.

If you are thinking of curing bacon as a commercial enterprise, a fridge is a necessary investment. It will save you a lot of time and energy, and you will have control of your product. Without a fridge, you are very limited.

Standard sweet pickle brine

Ingredients

> 12 gallons of water
>
> 22lb salt (ground finely)
>
> 6oz saltpetre
>
> 1-2lb sugar of your choice (can be increased or decreased according to personal taste)

Method

You will need a large container in which to put all the ingredients. The sweet pickle should come halfway up the container, leaving room for the pork. Make sure the container is scrupulously clean, then boil the water and pour into the container. Add salt and mix well. Heat sugar in another container making sure it is fully dissolved, then put in with the salt and water — stir well. If you are using a good quality brown sugar, the mix should be a lovely dark colour. Put saltpetre in a saucepan and boil until dissolved, stirring all the time. Wait until cooled to about 35°F, then leave for 24 hours. Finally, add it to the brine. The desired salt density (which can be tested with your brineometer) for the combined solution should be 70%.

Hanging the bacon

When drying the bacon it is best to tie a piece of good quality string round the hock of the ham and create a loop so that the ham is hanging straight.

With the middles you should thread the string through a corner, using a curer's threader (see picture below), so that the middle is hanging in a 'V' shape; this enables the moisture to drain off. With the shoulder, thread the string through the hock; this will allow it to drain. Thread the string in the middle of a streaky, so the streaky is hanging straight. These measures will give you good dry bacon.

Maturing shoulder pieces

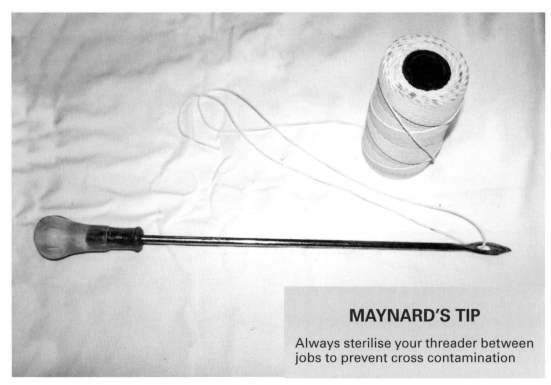

MAYNARD'S TIP

Always sterilise your threader between jobs to prevent cross contamination

The string threader and the special cotton string (avoid any plastic string as it will melt in the smoke house)

Always use a coding system for marking the bacon so that you know exactly the date on which the bacon was produced and by which method. Special marking pencils are available for this purpose.

Careful storage of your materials is also important — salt needs to be kept in a very dry place and your saltpetre needs to be kept in an airtight container because if it becomes damp it will go very hard. The herbs also need to be stored in airtight containers and in a dry area to preserve their flavours — make sure everything is well labelled.

A line of backs stored in a chilled room

Marcus Themans with one of his hams showing a nice white bloom

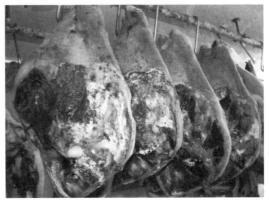

White bloom developing satisfactorily on a line of fine hams

Traditional Dry Curing

The wet curing method, where you make a brine as I have just discussed, is one way of curing bacon. A second method is dry curing, where you apply the salt, sugar and saltpetre directly onto the meat.

It is up to you which one you use as there are good points for both and it all depends on what kind of bacon you want to produce and for what kind of market.

The curing time will depend on the size of the cut: hams need the longest time, backs will take a little less, and streakys a little less still.

Today it is very rare that a whole side of pork is used for dry curing, but in days gone by it was done by laying down a bed of salt and putting the complete side on it, then inserting some of the dry curing mix in the blade bone before stacking them (rind side down) eight at a time, one on top of each other. These would be cured for 14 days.

You split the curing mix into two: for the first 7 days you use half the mix; then you take the sides out to re-salt them with the remainder of the mix. You then stack them the other way up — one to eight — rind down.

You then take the sides out, wash them all and hang them up in a cool place or in a fridge for maturing. The temperature should be in the range 40-42°F. This is the right temperature range for them to 'equalize.'

In other words, when the bacon comes out of the salt, the curing process is not complete. The salt, saltpetre and body fluids need further time to finish the process of equalizing. Once this is done, the colour and flavour will be fully developed and the longer you keep them hanging up the better the flavour. That is the process of dry curing by the traditional method.

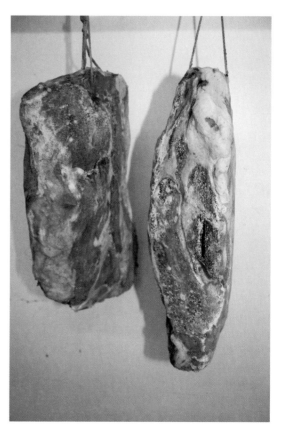

White mould developing nicely on a dry cure

42

Middles maturing in shallow trays

Prior to dry curing nowadays, it is more common to split the sides into smaller pieces. I prefer to split the side into three: ham, shoulder and middle.

After curing, I recommend that you dust the pieces with black pepper on both sides and put them in a calico or cotton bag to mature. To hang them, put string around the hocks and over the cotton bag. To string the middle, which needs to hang in a V-shape, use a curer's threader (see page 40).

To string the streaky, take the corner and string this: this will give you the V-shape. Leave now until the bacon becomes firm. It is better to use the shoulder first, then the middle, then the ham.

Dry Cured Middles of Bacon

For middles, apply the same principle as above. Put an inch bed of salt down in your curing container and then put one middle one way and the next middle the other way and build your stack into a pyramid, usually about six high and always rind down. The curing time for these is less than for whole sides, usually about 10-12 days. Judging the time for this comes with experience; the longer the bacon is left the saltier it will become. Turn after 5 days, top to bottom.

When you are dry curing and you have your pyramids of bacon, it is best to use a shallow container for curing with an outlet to drain the fluid. If you can keep the fluid flowing away from the bacon, this will ensure you end up with a dry product.

Dry Cured Back & Streaky Bacon

The process for the backs is the same as for the middles. Build them up like a pyramid, starting with five backs on the bottom, four on the next layer, three on the next, and so on until you end up with just one on the top. As the backs are smaller than the middles, less time is needed for curing. Streakys will take the least amount of time: less curing mixture is applied and the time taken for curing is about 5 days.

When all the bacon comes out, it is essential to wash it with cold water and to hang it in a cool dark place to mature. Light seems to affect the bacon detrimentally, and the temperature should be kept in the range 40-42°F, with a slow-running current of air. So hang them in a temperature-controlled area but not in a fridge. Remember to hang the backs and the streakys in a V-shape so that as they dry any moisture can run off them.

Traditional dry salted streaky bacon showing a firm, white fat

Traditional Dry Curing of Hams

These are cured in the same way as described earlier, but on a bed of ordinary (cheap) salt below the hams, with the dry mixture applied above. The bed of salt starts the osmosis process, drawing the moisture from the ham into the under-lying salt. As the salt below becomes wet, replace it with new dry salt.

Hams are cured over a period of 14 days, applying the first mixture and leaving for 7 days, then taking it out of the salt, removing the 'spent' salt (this is the salt that has gone wet) and re-applying with the fresh mixture and leaving for another 7 days.

After the 14 days, wash the hams with tepid water, string the hocks and hang up to mature until a bloom, a white mould appears on them. This is a perfectly normal white bloom — any other colour would indicate a problem.

Maturing middle and hams

The next step is to dry them fully and to apply an olive oil mix, which you make by warming 6oz of olive oil and adding 3oz of black pepper, freshly ground. You spread this olive oil mix evenly over the ham, and the easiest way to spread it is to use a good quality paintbrush. The oil will protect the ham and enhance the flavour.

Dry salted hams maturing

When fully matured you put the ham in a calico or muslin bag; the longer you keep the ham, the nicer the flavour.

Hams should be stored at a temperature of 45-50°F to mature. As they mature you will find the ham becomes harder and harder. To make sure your ham is sound, take a long needle, which has been sterilised, insert it into the bone joint, take it out and smell it. If the needle smells alright, the ham is alright. But always make sure you fill the hole where the needle has

Good white mould on a dry-cured ham

been with a piece of soft fat — this will seal the hole. For each new ham you test, a new sterile needle is essential.

So that is your traditional dry salting method. It is not a frequently used method nowadays as the bacon it produces has

a strong, salty flavour and this is not in favour today. But there are other methods of curing bacon and you can choose which is suitable for your own bacon production.

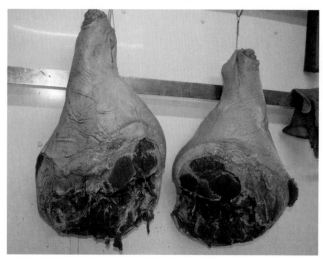

Maturing hams leaving plenty of space for circulating air

I have talked about wet cures and dry cures, and now I will go into a third method of curing. The one I am going to describe involves the use of a brine pump and for this method, which is known as a *combination cure*, you first have to know how to use a brine pump.

Artery Pumping

For this method of curing bacon, you will need a double-action brine pump and a straight needle with a single hole at the outlet.

The method is to insert the needle into the main artery in the ham, making sure that the artery has not been removed in the process of butchering.

As you pump the brine into the artery you will see the ham growing larger. The artery will distend, and fluid will begin to flush from the top of the ham.

Once you have reached this point, you will have completed the pumping and should stop. You want the brine at about 30-40% salt density — a brine of about 70% will make the product too salty.

The ham is not completely cured at this point and there are two ways of continuing, as follows. It is basically a matter of preference.

Combination: pumping plus wet cure

After artery pumping, to proceed with the combination wet cure, drop the ham into a sweet brine. If you put your ham into a sweet brine it will need leaving for about 5 days. This will give it a nice flavour.

After it has been in the brine for this time, remove it and wash with cold water. It is important to do this because if you leave any salt it will attract moisture and mould will then grow on it. So make sure this part of the procedure is not missed.

Now take some cotton string and string the pieces; then hang them up to dry and mature.

Combination: pumping plus dry cure

Another method of **combination** curing is to artery pump, then to apply a dry salt cure (as described previously).

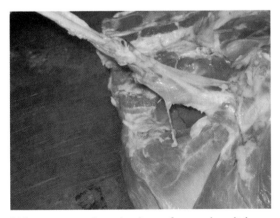

When preparing the ham for curing it is vital to preserve the femoral artery

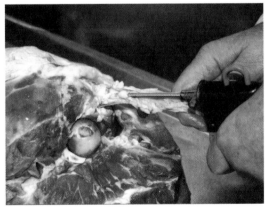

The artery needle is inserted and the pumping can begin

Pumping Middles

Middles of bacon (and streakys) can also be pumped. This is achieved by putting the middle on a flat surface, inserting a special needle and pumping at two-inch intervals along the whole length of the middle. (With the streaky, you inject about three times along the side and once at each end.)

The needle for this pumping technique (as opposed to artery pumping) is known as a 'brining needle.' It has several holes along the side of the shaft to facilitate distribution of the brine into the meat, and this technique is called a 'stitch'.

Once you have completed this, you use one of the two follow-up methods described, either putting them into a brine or dry salting them.

For the brine method, the middles will need to be left for 4-5 days; but for streakys, 3 days will be sufficient. Always use a fresh brine for pumping — never take it from the brine tank as it is not sterile and debris can accumulate in the tank so it would be most unsuitable for the pumping

method. It is best to make the brine well before you wish to use it as matured brine gives the bacon a tastier flavour.

You can put many different flavours into your pumping brine, thus giving many subtle flavours to your bacon.

But here again, I will repeat, never use the brine out of the tank for pumping; always make a fresh brine for this procedure. The reason for this is that a completely sterile brine is always needed for pumping. If this is not used you will run the risk of introducing harmful organisms into the very centre of your bacon, with disastrous consequences.

MAYNARD'S TIP

When you have finished using your brine pump make sure you run two buckets of water through it to remove any salt residue. Salt, if left, will corrode the machine. Dry thoroughly. Also, purchase a gallon of liquid paraffin: use this to oil all the working parts of your pump and all your needles and wrap the needles in a clean cloth.

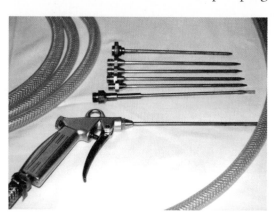

Brine gun and a selection of needles

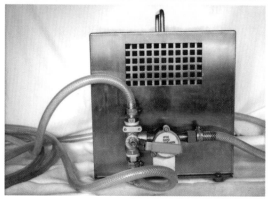

Maynard's own design of the brine injector pump, now sold by the Windsor Food Machinery

The automatic bacon injector – the right equipment for someone planning large-scale production

An old hand pump injector (there are modern plastic versions) – the better option for the sole trader or the small-scale producer, but not so efficient as an automatic brine pump

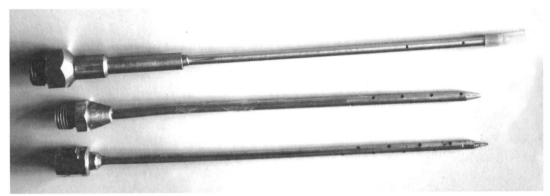

Three pumping needles - they can come with differing numbers and positions of holes, each one suitable for specific cuts of pork

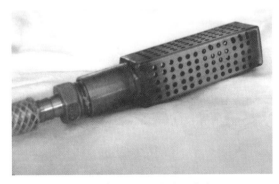

The filter (inset) for the brine injector

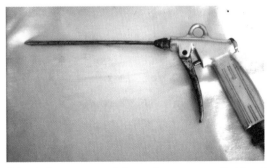

The injector gun (above), an essential tool for wet curing on any scale

COMBINATION CURE RECIPES

A combination cure is one that combines different curing methods, such as brine pumping with a follow-up dry or wet cure. As a general rule, the density for a pumping brine should be 30-40%. After pumping, the meat is given a dry cure or a wet cure for the usual time so that the cure is evenly distributed throughout the meat. If you decide to give it a wet cure, an ideal density for this brine would be 65-70%. Leave the meat in the brine tank for 3-4 days; then wash off in tepid water and leave to equalise.

Wet & Dry Salting Combination Cure

The cure I am going to describe now is a combination cure, but this one is a process of half sweet-pickling and half dry-salting. The southern half of the country favours this as it produces a gentle kind of bacon. For this recipe you will need to wash the bacon in a cleansing brine as previously described and then make yourself a brine consisting of:

> 10 gallons of water (make sure it is boiled and sterilised before using)
>
> 20lb salt
>
> 6lb sugar
>
> 6oz saltpetre

Boil the saltpetre in a saucepan until dissolved, using a small amount of water. Boil the sugar in enough water to dissolve the sugar, and leave to cool. Mix the salt and water, making sure the salt is dissolved, and then add the dissolved saltpetre. Leave to go cold; then stir thoroughly. Add the sugar (when cooled) and thoroughly mix, making sure all the ingredients are dissolved.

The shade of the pickle should be a golden colour. The next step is to measure the density of the brine, using your brineometer. Sterilise a bucket, dip it into the pickle and take half a bucketful out. Drop your brineometer in and test the density. This is important as the strength of the brine should be about 65-70%. This will give you a nice gentle pickle.

Put all the bacon into this pickle and leave for 5 days, remove and dry off thoroughly. Now you will need to salt the bacon, using the following mixture:

> 24lb salt
>
> 8oz saltpetre
>
> 6lb Demerara sugar

Mix the ingredients together and lay the middles down first. Salt them using the mixture, applying 1oz per pound of meat. Now lay the shoulders, and then the hams, salting each with a good layer of the mixture on the top. Leave for 6 days before breaking the pack; then wash them off and dry thoroughly. Put into a muslin bag and leave in an airy place to mature.

Brine Pumping & Dry Salting Combination Cure

This is a combination cure where, first of all, you inject pumping brine into the meat. This hastens the curing process and restricts any spoilage as the cure is taken straight to the centre of the meat. To achieve this, you inject the meat with brine using the 'stitch' method. When you are injecting a middle, you usually inject at 2" intervals along the middle. With the streaky, you inject about three times along the side and once at each end. Do not over-pump as you will destroy the tissues of the meat. Just pump gently, and the density should also be mild — about 30-40%, and no higher than that or you will find the end result will be too salty.

When you have injected them, you then dry salt them with a dry salting mix. I recommend the following mixture, but you may want to use a mix of your own choice:

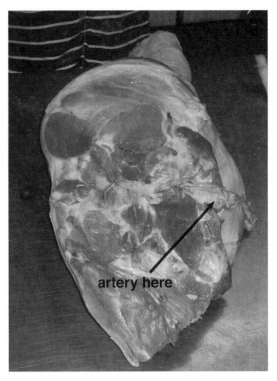

Long cut ham ready for curing and showing the position of the femoral artery used for pumping

> 25lb large grain rock salt
>
> 8oz saltpetre
>
> 6lb Demerara sugar or light Muscovado sugar

Mix these ingredients thoroughly, then store in an airtight container, making sure the lid and sides are well marked with the ingredients — this is your dry mix for the combination cure.

The meat is laid (rind down) in a container with a hole in the bottom. Sprinkle it well with your mixture, 1oz per pound of dry salted meat. The middles and the streakys will take about a week to cure. The hams will take a lot longer — roughly about a fortnight.

MAYNARD'S TIP

Pumping has many advantages as a way of curing. It takes the brine into the centre of the meat and so prevents spoilage. Also, you can decide which flavour you want the meat to have. Curing time is reduced and wastage is reduced. I recommend this way of curing.

Maintenance of Brine Tanks

When your bacon has been removed from the brine tank, maintenance of the brine is essential if you want it to carry on. You must have a sieve and remove all meat particles from the brine. This is essential, for the brine will go sour if pieces of meat are left in the container. If the brine is well looked after and filtered thoroughly and regularly, it will last for a considerable amount of time, sometimes for years.

To restore your brine add salt, saltpetre and sugar when necessary — but bear in mind you may be altering the salinity so check it with your brineometer. It is a good idea to have the brine tested by the Public Analyst every month, and then you will know you are complying with Food Regulations.

Brine will last for a long period if it is well maintained and kept at the right temperature. If not, your brine will go ropey and that means it will go very thick or cloudy. Some Curers will take it out and re-boil it but I do not think this is a good practice and I personally favour getting rid of it and starting again.

One of my methods of prolonging the life of brine is to transfer old, mature brine to a clean container by using a submersible pump. The sieve on the pump will remove all the impurities, leaving the brine a nice claret wine colour. It is a fact that the longer the brine is kept the better the flavour becomes.

After you have mixed a new brine the density must always be tested with a brineometer. To do this, take a sample from the container in a clean sterile stainless steel bucket, mixing well before you do this, wait for the mixture to settle and then place the brineometer in the bucket. The higher the reading, the more salty the product (see Appendix).

Brine should be kept as long as possible as the flavour improves. I used to keep mine for years and it was always a very sad occasion when a brine had to be thrown away; it was like losing an old friend!

Tumbling

Tumbling is an ideal method of distributing the cure and is used by all major food companies. The way it works is like this. The modern tumbler is like a large barrel with steps or ledges in the side of it. Meat for curing is put into this round container, and the meat hits the ledges and tumbles about.

In the industry, people think this is a new invention but it is not. Many years ago the old Curers used to put meat into barrels and roll them up and down the curing cellars. In effect this massages the meat and facilitates the osmosis of the brine. It was done so that the meat would have a more even cure and a more uniform colour, with no paler, poorly cured patches.

You can sometimes see in delicatessen shops hams where there are two colours. Tumbling helps to get rid of that and so you get a better presentation. You also get a softer product.

There is another advantage to using a tumbler. Curing time is reduced as you get a faster penetration of the cure. This means you can produce and sell a greater quantity of produce in a shorter time. Tumbling is ideal for hams, gammons, silversides and topsides, which are best presented with a uniform colour.

The best way to use tumbling I found was firstly to inject the article you intend to tumble and give it the correct amount of time in cure, fetch it out and wash off. Now put it in the tumbler, and this will give the article a uniform colour.

The brine you intend to put in the tumbler is best left in a bucket in the fridge overnight, so it is very cold. You will find that tumbling the meat will raise the temperature of the meat.

To offset this, if you put cold brine into the tumbler you will find it will keep the temperature of the meat down.

Some manufacturers add polyphosphates to the brine they put in the tumbler (see Chapter 3), but I have never done that; I used to tumble with a simple brine.

A tumbler is a piece of machinery, but it is only for those of you who want to specialise in the production of hams, topsides of beef, silversides, and to a degree pastrami. It is a good piece of machinery and I thought it would be a little piece of information for those of you who want to go down that road.

Thin slices of bresaola

Applying the coating to Parma hams in the traditional Italian manner

CHAPTER FIVE

CURING RECIPES

The recipes in this book are traditional recipes. Many were used by our ancestors, and some are very old. In the past, the main objective of curing was to make sure that our ancestors had bacon over the winter months, and the bacon was salted so that it would last. Salt was the main ingredient and we must always remember that it was used in large quantities so that it would act as a preservative and keep the meat for long periods. The harsh, salty bacon of yester-year is not in favour today; we prefer a less salty kind of bacon. But for posterity I have

included these recipes because they are your heritage and you should know how bacon was once produced. In days gone by, salt was only added to cheese, butter and bacon. But today salt is added to most food products and this means that heavily salted bacon is not popular or necessary. For the Curers of today, it is a good idea to have these recipes to hand as you never know when you may be called upon to produce a traditional English bacon. I will begin this chapter by describing some dry cure recipes.

A plate of air dried ham

RECIPES FOR TRADITIONAL DRY CURES

Roman Ham

This is the oldest recipe I have used. It is an original Roman ham recipe and I estimate it is about 2,000 years old.

Ingredients

> 14lb Roman salt
>
> 4oz coriander (dried)
>
> 1 pint good Italian red wine
>
> 6oz Italian honey
>
> 4oz olive oil
>
> 4 fl oz of white wine vinegar

Method

Wash hams in a cleansing brine, and dry off thoroughly. Mix together the salt and coriander. Put a scattering of salt on the bottom of a container, making sure that the container has a hole in the bottom to drain off excess fluid.

Put ham in the container and scatter the dry salt and coriander mix over it. Repeat with other hams but only stack two hams high; if you go too high, the juices in the hams will be forced out. Make sure the hams are skin down in the container and cover the hams completely with the dry mixture.

Leave them like this for 5 days, after which you should rotate the hams from top to bottom, re-salt and make the salt level so that none of the hams are uncovered and visible. Leave for a further 12 days, making a total of 17 days.

After the 17 days, break open the salt, take out the hams, and wash them in cold water. Then tie a piece of good quality cotton string around them, and hang them up to dry for about 4 days. When thoroughly dry wipe them with a clean cloth.

Take the red wine and honey and mix together thoroughly in a clean container. Use a sponge or paintbrush to paint this mixture on the hams, taking special care to coat each ham fully. Leave to dry for three hours, and the hams are now ready for smoking.

To protect the hams from flies and pests, I recommend that you wrap them in muslin cloths, but this is optional. Hang them in the smoke house and smoke until they develop a golden colour.

Leave the hams in the smoke house to cool down, then remove and take them to a cool area making sure the hams are

A dry cure mix (plastic is the best material for the container). An even distribution of the ingredients is the key here

completely cold. Remove the muslin cloths. Paint with a mixture of olive oil and vinegar. When dry put them into a calico or cotton bag and leave to mature. This is the oldest recipe I have come across and you will find it is an excellent ham.

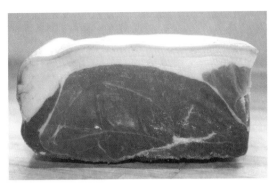

Dry salted ham

MAYNARD'S TIP

Remove bones before curing as mixtures will penetrate the meat much better.

Melvyn Ling of Appleyards, Shrewsbury, with some of his maturing dry-cured hams

Traditional Ayrshire Bacon

This is the old Scottish recipe, which in my opinion makes really flavoursome bacon and is the dry cure version. You will need a pig that is not too fat — a lean pig.

Make sure the meat is thoroughly cleansed with a cleansing brine and has been refrigerated overnight.

Take all the bones out of the middles, making sure you do not cut too deeply into the meat. Take the long trotters off the hams so that they are a little shorter. Take the bones out of the shoulder, remove the rind and thoroughly dry.

Now make the curing mix (this is another dry cure):

25lb salt

6oz saltpetre

2lb Demerara sugar

5oz pepper

Make a bed of salt in the container and put the first middle in. Scatter some of the curing mix evenly over the middle, about ½oz per pound of meat. Then put the other middle the opposite way, and skin down so that the two middles fit snugly into one another like a glove.

Now put the shoulders together and put about 2oz of the cure in the top of the shoulders (where the blade bone has been taken out). This cure will go straight into the shoulder. Place the hams opposite to each other, sprinkle the cure over them and leave for about 5 days.

Remove all the cuts from the cure and put them back in opposite directions, swapping them over so they are all in new posi-

tions. Replenish the cure if you need to.

It is a good idea to take the middles out after about 8 days, wash them, dry them and then put a muslin cloth on them and leave for a further 5 days. Re-salt the shoulders and the hams and leave them for a further 8-10 days, as the hams and shoulders will take longer to cure.

Once the middles have been in the cure for sufficient time, remove the muslin and take off the rind with a very sharp knife. Now take a meat hammer, which is a very long piece of wood like a cricket bat, and hit the middle. You will find that the middle will spread and become level.

Now for the seasoning. Take 1oz nutmeg, 1oz mace and 1oz coriander. Mix them together and put in a shaker. Spread this mixture in a thin layer across the middle, starting with the thick side. Roll up the middle and the streaky in a long roll, then put a string around the middle of the middle.

Now tie another string on one side and then another on the other side, working from the middle out. The strings should be about an inch apart and make sure they are very tight. Then put the whole roll back into a net to mature for about 3 or 4 days. This will then produce a spiced bacon roll which, when matured, can be cut very thinly. It can also be smoked as a further process.

MAYNARD'S TIP

Stringing from the centre and working towards the outside edges produces an even roll.

A dry-salted, long-cut ham ready to be placed on the bed of coarse sea salt. Ideally the container will be made of plastic, to avoid long-term corrosion

Derbyshire Favourite

This is an old-fashioned recipe. Take your pig and split it into shoulders, backs, streaky and hams, and put the pieces in a cleansing pickle for the recommended time. Remove and dry, placing them rind down. Now mix together:

24lb large grain rock salt

7oz saltpetre, finely ground

4oz dark pepper, finely ground

6oz juniper berries, finely ground

Mix all these ingredients thoroughly. Make a bed of rock salt, about an inch thick, in a container. Then put the middle on the bed, rind down, and cover with the mixture, about ½oz to 1lb of meat. Then put the back the other way and build up a pyramid, salting with the mixture as you build. Then add the shoulders and finally the hams on the top, covering them with the dry mix. Leave for 5 days before taking the streakys out. Wash and dry the streakys and put them in a muslin cloth. Leave the backs, shoulders and hams in the salt for another 5 days. Then take them out, wash and string them (as described previously), and hang them up to mature. This is a strong kind of bacon.

MAYNARD'S TIP

Mark the bacon with the date of its curing using a food industry marker pen.

Farmer's Sweet Cure

I have used this old recipe for many years.

Ingredients

48lb fine salt (i.e. salt, ground finely)

9lb dark Muscovado sugar

9lb Demerara sugar

14oz saltpetre, finely crushed

3lb sea salt (large grain)

Method

Blend all the ingredients together making sure they are well mixed. Be sure there are no lumps in the sugar as it is essential there is an even distribution of the ingredients. Take the ham, back and streaky and put them in a cleansing brine to make sure all the debris is washed away. Remove and thoroughly dry.

Put an inch of salt in a container and start to build your stack. Put the backs in first — one back one way and the other back the other way — then the streaky, then the shoulders. In the 'pocket' of the shoulders, where the blade bone has been removed, place about 2-3oz of the cure. Finally, add the hams: one in one way and the next the other way. Once your stack has been made, cover it completely with the cure, making sure there is no meat uncovered.

Leave the pyramid in cure for 5 days before breaking the pack and taking the streakys out. Return the hams, backs and shoulders to the cure, placing them tops to bottoms as previously described, and after 3 days remove the backs. Leave the shoulders and the hams for another 2 days. Thus the respective times in the cure are:

streakys — 5 days

backs — 8 days

hams and shoulders — 10 days

When you take out the pieces of meat, wash them thoroughly with tepid water — this will release all the salt. Then dry and string them. Put them in a polythene bag or muslin to mature them, but you *must* dry them thoroughly first.

Spicy Bacon 1

This recipe is for a spicy bacon, and it works better using a lean pig rather than a fat pig.

Ingredients

35lb rock salt, finely ground

2lb sea salt

10lb Demerara sugar

14oz saltpetre

4oz coriander

4oz black Jamaican pepper

Method

Put all the fine salt into a container, plus your sea salt followed by your sugars. Make sure your saltpetre is finely ground and distributed well. Then add the pepper and the coriander and mix all the ingredients thoroughly. Remove the bones from the middles and the backs. Place middles in first followed by the streakys, then the shoulders and then the hams on the top as described previously. Remember to apply the mixture between each layer and make sure there is a good layer on the top, making sure no pork is showing through.

The cure should take between 10 and 14 days. If you do not want your streakys to be too strong, take them out halfway through the process and wash them off. This recipe produces a delicious spicy bacon and it should give you a lot of pleasure curing it. This is a well-proven recipe and the proportion of cure to meat is: ½oz of mix to 1lb of meat.

I used to find this kind of bacon nice to smoke, especially with a mixture (half and half) of oak sawdust and beech sawdust. This seemed to give it a lovely flavour and to enhance it further I would throw some juniper berries in the smoke — this would give the bacon a lovely aroma.

MAYNARD'S TIP

Keep your spicy bacon mix in a container with a tight-fitting lid, putting cling film under the lid as a seal. This helps to keep the aroma in and retain its freshness.

Spicy Bacon 2

Ingredients

16lb large grain rock salt

16lb Bay salt

6lb Muscovado sugar

12oz saltpetre

6oz Jamaican pepper

6oz ginger

Method

Mix ingredients thoroughly, as previously described. Apply the mixture, ½oz to 1lb. This spicy bacon recipe works a little better if the bacon is on the fat side; it seems to take to the fat well and produces a lovely flavour with a hint of ginger. The cure takes about 8 days and it works better if you just use the middles, wash them off and then allow them to mature.

Lardons - see recipe on right

Lardons

The best cut for lardons is belly pork. Thick diameter belly pork is the choice of cut for this recipe; a good covering of fat with the rind on. Cut the belly into squares of equal size and this is now ready for curing. Wash the belly pork thoroughly in cleansing brine and dry in the usual way before starting to cure.

Ingredients

10lb rock salt (large granules)

4oz saltpetre

2oz white pepper

2oz mace

3oz garlic

6oz Demerara sugar

4oz juniper berries

Method

Take the saltpetre, pepper, mace and juniper berries; put them in a grinder and grind finely. Add this mix to the salt and sugar, but at this stage leave the garlic out. Put a layer of salt in a container, making sure the container has a hole in the bottom to release the excess fluid. Apply half the mixture (at a proportion of 1oz per pound of meat). After 3 days, take them out and re-salt them with the remaining mixture. Mix well and spread the second part of the mixture over all the belly pork.

Leave this for another 3 days, wash off with cold water and dry thoroughly. When dry apply the garlic. When the garlic has dried put the belly cuts into nets to mature. The longer you leave them to equalise the better the flavour will be. If you want a stronger flavour leave them in the cure for at least 6 days. Once taken out of cure remove the rind. Do not remove the rind until the cure is complete; this helps to keep shape and condition.

These bellies can be smoked if so desired, or left unsmoked. If smoking, put the bellies in a muslin cloth — this stops the smoke being too strong. When they are fully matured, and if you do not want to use them straight away, put the bellies in a vacuum bag and keep them in the fridge, making sure you date each bag, and fetch them out as required.

Lardons may be cubed or thinly sliced, whichever way you want to use them. They are an excellent product and you will find they will sell well for you. If even more flavour is desired, coat them with ground black peppercorns. Hang up only in curing nets and do not put a hook through them, as this will mark the meat.

Dry cured streakys and a row of legs just visible on the lower rack

Fat streaky: perfect for lardons

MAYNARD'S TIP

Vacuum pack your lardons, date them and use in rotation. If left out for too long they seem to fade, so vacuum packing extends the shelf life. Store in the fridge.

Air dried ham

Air Dried Ham 1

This old recipe involves maturing the ham for a long time and eating it raw and very thinly sliced.

Ingredients

    20lb rock salt
    6oz saltpetre
    2oz sage
    3oz bay leaves
    1½oz garlic
    6oz juniper berries
    3oz black pepper
    2oz nutmeg
    1oz chives
    2oz thyme
    4oz coriander

Method

Grind all the herbs. Take 6oz saltpetre and mix well with the 20lb of salt, then add the herbs and spices and mix them all thoroughly.

Wash the hams and dry thoroughly. Put a bed of salt in a container and lay the hams down, making sure the container has a hole in the bottom to let excess fluid run off. Cover completely with the ingredients and leave until it is a good colour.

After a week, break the pack, re-salt, and place the hams skin down. Cover completely with salt and leave for one month.

After a month take them out, wash them thoroughly and tie a piece of good quality cotton twine around the hocks. Hang them up to mature for a week. After a week, take them down and insert a long needle into each ham. Withdraw the needle and smell it. The aroma should smell fresh. If it does, take a small piece of fat, cover it with pepper and block the hole up with it: this prevents bacteria invading the ham.

The next thing to do is to put the hams in the smoke house and smoke them, using beech, for about a week. Take out, cool, and put them in a cotton bag to mature. These hams can keep for twelve months, but most people probably use them before that. They must be boned out very carefully, and left under a press for a fortnight to compact the shape. The hams should be sliced very thinly and you will find the taste is delicious.

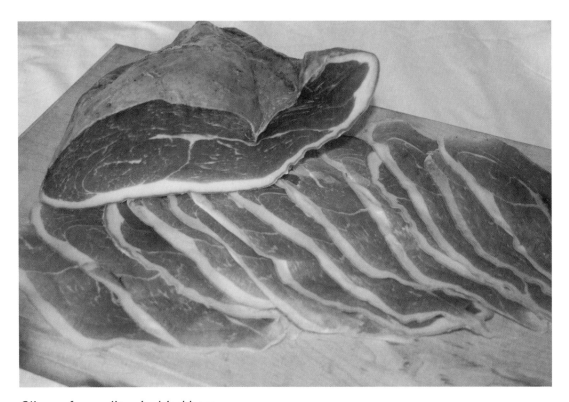

Slices of a quality air-dried ham

Air Dried Ham 2

This ham is for eating raw and it should be cut very thinly. You will need a 'short cut' leg of pork for this recipe.

Ingredients

30lb bay salt

6oz saltpetre

4oz black pepper

6oz coriander

1 pint of white wine vinegar

1 clove of garlic

Method

First, cover the hams completely with a layer of salt for 3 days. Once all the body fluids have been removed, apply the curing mixture.

For the curing mixture, mix together thoroughly the saltpetre, salt, black pepper, and coriander. Bone out the leg of pork, and then apply the ingredients inside and out. Where you have taken the bone out, put some of the ingredients in, using 1oz to 1lb, and then sew up near the hock to give a good shape to the ham.

Once you have put the mixture inside the leg of pork, put the leg in a curing net to keep the shape. Do not hang it up but lay on a rack so that it does not lose its shape.

Cure for 10 days; then fetch out and wash off. Now apply the garlic and wine vinegar to the dry ham and put under a wooden press. Press into shape, and leave for another week or until it is completely hard. Take out the ham, rub with garlic and vinegar, and put in a muslin cloth.

Hang up in an airy place and leave for about a month. You will find this is sufficient time to produce a delightful flavour. Cut finely on your machine and you will find this a very enjoyable ham.

When you fetch the ham out of the muslin cloth, you may find there is a mould on it. Do not be alarmed: if it is a black mould be suspicious, but if it is a white mould this is quite acceptable. You will need to rub it off with some more vinegar and you will find it will come off quite easily. Wrap again in a muslin cloth and you will find that you have an excellent product.

A selection of air dried hams and salamis

Fat bacon – still favoured in some parts of the country, especially by the older generation

Strip of fat bacon ready for cutting into larding strings

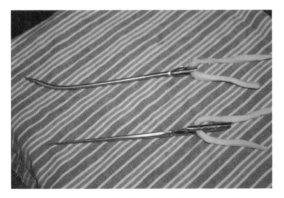

Fat bacon threaded on larding needles

Fat Bacon for Larding

Fat bacon (or fat pork) is all fat and is used for adding flavour and succulence to pork in roasting. When prepared and ready for use, it is cut into thin strips and threaded through a pork joint, using a needle known as a larding needle.

Ingredients

Large pieces of fat pork with rind on

28lb large grain sea salt

6oz caraway seeds

6lb sugar (Demerara or white – do not use dark sugar, as it will stain the meat when roasting)

4oz white pepper

Method

Put the salt in a large container. Grind the pepper and the caraway seeds very finely and mix with the sugar. Place the pieces of fat pork rind down as follows, applying the mixture to each piece: three pieces on one side of the container, three on the other, and stacked no higher than three. Leave them in the cure for 10 days, remove, wash off and dry thoroughly. Spread pepper all over, and put in a square container until fully matured. Fat bacon is not hung as it would lose its shape. Date and vacuum pack the pieces, put them in the fridge and you now have your larding bacon ready for use.

When larding pork, it is better to take the rind off the bacon prior to use and cut the tough fat into very thin strips before threading with your larding needle. Always sterilise the needle prior to use.

Warm the needle prior to insertion and you will find it will slip in easier.

The cuts that are best for larding are: fillet steak, tenderloin of pork, roast pork, loin of pork, chicken, turkey, and veal. Any meat that has a low fat content would taste more succulent with the addition of larding bacon and there is a vast range of uses for it.

Note that this recipe does not use saltpetre. This is because there are two ways of curing fat bacon. The one I have just described is a salt cure which produces a larding bacon with little colour and is good for tenderloin and light pork. The other method requires the addition of saltpetre. The saltpetre variation is more pink-coloured and might be used for a cured ham. So if you want to lard hams or bacon or any cured product, you must add saltpetre to the ingredients. But only use saltpetre in **cured** products. Do not use it in other products as the meat will be stained by the saltpetre. Saltpetre is not used on fat bacon for light meat because of the effect on colouring.

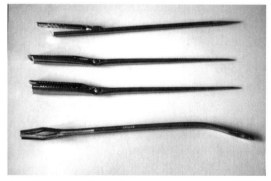

Larding needles of various sizes

Sweet Cured Bacon

This is an unusual dry cure recipe, which will produce a bacon with a distinctive, slightly sweet flavour.

Ingredients

    28lb salt

    6oz caraway seed

    6lb Demerara sugar

    4oz black pepper

    6oz saltpetre

Method

Put the salt in a container, grind the caraway seeds and the black pepper, add to the container and add the sugar and saltpetre. Mix thoroughly. Clean the meat in cleansing brine, dry off thoroughly, and put a bed of salt in a container. Layer the pork and cover with the mixture (as previously described) and leave for 5 days. Take the pieces out, re-salt them, and leave for another 5 days. Then remove from the container, wash them off, and make sure they are thoroughly dry. Cover with peppercorns and leave them to mature. You will find that the longer you mature the pieces (at 45-50°F), the more they will develop a lovely flavour.

MAYNARD'S TIP

Mark the bacon with the date of its curing using a food industry marker pen.

Romany Bacon

This was produced for many years for the Romany gypsies. I used to know the people who supplied all the Romany people in Britain with this bacon. It was a very hard, fat bacon, but they liked this bacon as it lasted in their caravans and it gave a good strong flavour to their food. It would last for twelve months or more. This recipe shows how it was traditionally cured.

Ingredients

40lb large grain rock salt

8oz saltpetre

6oz caraway seeds

8oz red pepper

15 bay leaves

6oz mace

20 juniper berries

Method

The ideal bacon for this recipe was very fat middles. They were covered with rock salt and left for a week. On top of the rock salt a wooden plank was laid, and on top of this large weights were put to weight it all down. This served to remove all the body fluids completely. After a week the pack was broken up, the middles were dusted off, and the ingredients were applied. The caraway seeds and peppercorns were ground, plus the bay leaves and the juniper berries. This was all mixed with salt and saltpetre.

Next, the middles were laid rind down on an inch bed of salt. Three middles were laid rind down and then another three middles on the top, adding more of the mixture to the top layer. The stack was three high and they would be left in salt for about a week. The pack was then broken again, all the salt dusted off, and the pack was re-ordered; then the re-salting process was repeated for a third time, and the pack was left this time for three weeks.

After this, the pack was broken up and all the salt washed off; then the bacon was spread with peppercorns on the rind and fat, and put into muslin bags to mature. After another three weeks the bacon was exceptionally hard, and that is how the Romany liked it. It was a delicious bacon and I hope that one day, if you ever have the time to do this recipe, you too will find it delicious. You could supply local hotels with this bacon to introduce a different flavour to the food.

Maynard's Apprentice Bacon

This is one of the main recipes I learnt when serving as an apprentice.

Ingredients

> 48lb sea salt
>
> 9lb dark Muscovado sugar
>
> 9lb Demerara sugar
>
> 14oz saltpetre

Method

Cut your pig into sections: middles, shoulders, and hams. If you want to, you can make picnic hams out of the front end.

Next, put the meat into a cleansing pickle and remove once the meat has been thoroughly cleansed. Now put on a rack, rind up, to drain thoroughly.

When drained and dry, put the meat in a curing container, rind side down, and cover completely with salt. Leave for a day before removing.

Now dust the salt off and apply the curing mix — ½oz of curing mix to 1lb of meat. Lay a bed of salt in the container and put the middles down first, then the shoulders and then the hams on the top. Leave them in the cure for 7 days.

Now break the pack and re-salt for another 7 days. If you do not want a strong bacon, you can reduce the amount of time in salt. After breaking the pack, wash off the pieces in cold water and hang up to dry.

When dry, put some olive oil on the middles and the hams, and put them in a cotton bag to mature. When matured,

add peppercorns for more flavour. This is a long-keeping bacon, and the choice is yours as to whether you want a strong bacon or a mild one. Do not smoke this bacon as it will be too harsh.

Ham and bacon in muslin netting

MAYNARD'S TIP

The hams will need maturing at a temperature range of 40-50°F. Do not put them in a fridge but mature in a dry airy place. These hams need a longer, steadier period to mature and you will find the aroma will be lovely.

Dry Cured Roast Ham

I used to produce this ham for special occasions: christenings, weddings, any special party or large gathering. It pleases many and disappoints few.

Ingredients

35lb rock salt

2lb sea salt

10lb Demerara sugar

14oz saltpetre

4oz coriander

4oz black pepper

Method

Mix the sea salt, rock salt and Demerara sugar thoroughly. Mix the saltpetre, coriander and pepper and stir well. Add this mixture to the sugar and salt, then put in a container for storage. Lay a bed of salt in a container and place the ham on top. Apply half the above mixture to the ham, making sure it is well covered, and leave for 4 days.

Remove from the container, break up the pack and re-salt for another 4 days (making a total of 8 days in cure). Remove from the container, wash off and dry thoroughly.

Now leave them to mature, first placing them in a calico bag or a muslin cloth. These hams need a long maturing period so that the ingredients can equalise and for the full flavour to be achieved — allow 2-3 weeks.

When mature, the ham is ready for cooking. The night before, soak the ham in water with a little honey added. Change the water after a couple of hours and soak until the following morning. Remove, dry thoroughly, and remove the bone.

Attach a piece of string to the bone and re-insert in the ham. The reason for doing this is that the ham will keep its shape and the heat bounces back off the bone and the ham cooks better.

At the end of cooking and cooling, all you have to do is pull the piece of string and the bone pops out, leaving you to carve the ham easily and making a nice shape for display.

Now put your ham in a container and roast for the prescribed time, 20 minutes to the pound. Use a basting juice (as described on pages 152-155). Once cooked, remove from the oven, remove the rind and leave to cool. When cooled, wrap in a clean cloth until ready for use.

This is an excellent ham for functions and special occasions. It is not as spicy as other hams as it has to go with other ingredients at the functions.

MAYNARD'S TIP

Do not keep your hams in the fridge but hang them in an airy place where it is cool and dry. Cover with a muslin cloth.

Penitentiary Dry Cured Bacon

This recipe is one I used when I worked in the State Penitentiary in Pennsylvania (I worked there, by the way – I was not an inmate!) The inmates ate a lot of greens and vegetables and this bacon was meant to give their food a bit more flavour. The bacon was a fat bacon as the hogs were fed on corn, vegetables and greens — they were very fat pigs.

Ingredients

> 16lb fine salt
>
> 16lb bay salt
>
> 6lb dark Muscovado sugar
>
> 12oz saltpetre
>
> 6oz dark Jamaican pepper
>
> 6oz ginger

Method

In the States we used to cure in a fridge at about 42°F, as the temperature outside could be very high indeed. Also, we cured all the year round so we needed a stable temperature. We would mix all the ingredients in a large container, then apply them to the meat, 1oz to 1lb. We cured the bacon for about 12 days.

After 6 days we broke the pack, re-salted the pieces, and then put them back for another 6 days. Then we removed them, washed them off and dried them. We would string them up and make sure they were thoroughly dry before leaving them to mature.

Once matured, they were put into the kitchen production. They made some nice

tasty dishes which made just that little bit of difference and also helped to make the prison self-sufficient.

MAYNARD'S TIP

When using the spices, only sprinkle lightly and rub in well. Do not overdo spices: you can always add but you can't take away.

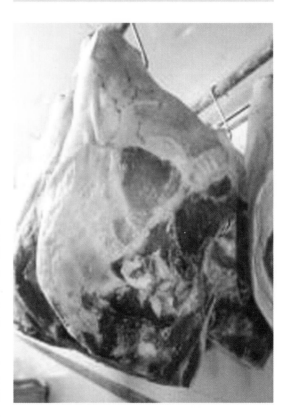

RECIPES FOR WET CURES

Now I will turn to the wet cures (sometimes known as sweet cures) which are very interesting and can be used to enhance the flavour of your bacon.

In my opinion, it is easier to produce a good brine-cured bacon than a dry-cured ham as you seem to have more control over it.

A brine cure generally involves putting the meat into a brine tank and leaving it for 5 days; then it is removed, washed off, dried thoroughly and left to equalise.

Ready for immersion in a sweet brine

MAYNARD'S TIP

When you have taken your meat out of a brine you should always sieve the brine to remove any remaining bits. Brines will keep for a long period provided there is no debris left in them.

Old Style Sweet Brine

Ingredients

40 gallons of water

50lb rock salt

1lb saltpetre

3lb bay salt

4lb Muscovado sugar

1oz coriander

2lb honey

Method

Put the water in a container. Boil, add the rock salt, and stir thoroughly. Dissolve the saltpetre in a saucepan, wait until cooled and then add to the brine. Add the Bay salt and stir well.

Dissolve the Muscovado sugar with a small amount of water in a saucepan and add to the brine when cold. Grind the coriander, boil with a small amount of water and add this to the brine.

Put the honey in a saucepan with a small quantity of water, dissolve and add to the brine when cold. The density of the brine should be 65-70%, and I recommend a pumping density of 55%.

American Breakfast Bacon

Ingredients

8 gallons of water

16lb bay salt

4lb molasses sugar

2lb treacle

5oz saltpetre

Method

Boil the water and put it into a container. Grind the saltpetre into a fine powder. Using the saucepan that you keep for saltpetre, dissolve the powder in a little water, boil and leave to go cold. Now, in a separate saucepan, boil the treacle and add a little water to thin it. Add the sugar and boil until dissolved. Now put all ingredients into the water and stir thoroughly, making sure everything is dissolved. The density of this brine should be 65-70%.

Remove your meat from the fridge. Remove some of the brine and dilute to a density of 35-40%. Add this brine to your pumping machine and inject each piece of meat as described previously. Once pumped, the decision now is whether to dry salt the pieces or to put them into the brine.

If you decide to put them into the brine, make sure you record the time and date. If you want a gentle bacon, leave them in for 5 days — this applies to the middles, shoulders and hams. The streakys should come out in 3 days. Once the meat has been put in the brine, make sure you put a weight over it to keep it submerged, and after 2 days move the pieces around so they are all in different positions. This

MAYNARD'S TIP

After a brine has been used, take a sterile bucket and remove some of it. Test with a brineometer and make the brine up to the correct density. It is then ready for use at a later date.

is known as 'overhauling'. Make sure the weight is returned to the top.

Once the appropriate time has been reached, remove the meat from the brine, wash off with tepid water, dry and hang up. If you want a stronger bacon, keep the meat in the brine using the formula: one day for every pound. Wash off the pieces, dry them and leave to mature. You will find this bacon will be a little darker than average because of the treacle, but it will give you a lovely taste and makes an excellent breakfast bacon.

Maynard with Neil Hollingsworth of the Dukeshill Ham Company with his black Shropshire ham

Spicy Brine 1

This is a brine I used for tenderloins and streakys.

Ingredients

10 gallons of water

20lb salt (I recommend rock salt or Bay)

6oz saltpetre

1oz allspice

1oz black pepper

2oz juniper berries

1oz coriander

2lb Muscovado sugar

1 pint of treacle

1 quart of strong beer

Method

Boil the water and put into a container. Add the salt and the beer. Grind the allspice, pepper, juniper berries and coriander. Put them in a saucepan with a little water, boil and wait until cooled. Then add to the container.

Boil the sugar in a little water and then add this. Dilute the treacle with a small amount of warm water. Wait for this to cool and then add to the mixture. The saltpetre should be dissolved as previously described and added to the mix.

Once all the ingredients have been mixed well, you are looking for a density of about 70% on the brineometer. This will give you a nice spicy bacon.

This brine is ideal for tenderloins and streaky bacon. With tenderloins, do not leave in for any length of time — 2-3 hours is plenty. The streakys will want about 3 days. When cured, remove them and dry off. You then have two options with streakys: you can either smoke them or leave them unsmoked.

If you decide to smoke them, smoke them in a net: you do not want too heavy a smoke with this recipe. You can also smoke the tenderloins — dry them off and smoke them in a net, and you will find you have an excellent product.

Hams curing in stacks of plastic curing trays

Black Country Ham

Ingredients

> 10 gallons of water
>
> 20lb sea salt
>
> 6oz saltpetre
>
> 2lb Bay salt
>
> 6oz juniper berries
>
> 6oz coriander
>
> 2lb treacle

Method

Boil the water and put into a container. Put in the sea salt and stir well. Dissolve the saltpetre and boil as in the previous recipe; leave to cool then add it to the brine.

Take the Bay salt, add it to the brine and stir well. Take your juniper berries and coriander, put into a grinder, blend well and add to the brine.

Put the treacle into a large saucepan, add a small quantity of water and heat until dissolved. Stir and make sure it is mixed thoroughly. When cold, add this to the brine.

Using your brineometer, measure the density: it should be about 65-70%. (If the meat is to be pumped, pump at a density of about 35-40% — this will give you a good ham.) Cure for 5 days.

Remove the meat, wash off, dry, and hang in the smoke house overnight to make sure the pieces are thoroughly dry before smoking.

Wrap the ham in a muslin cloth, and smoke using half beech and half apple wood. This should take about 2 days and

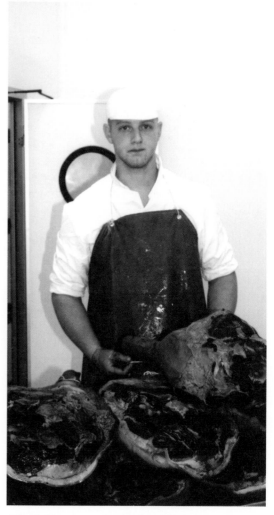

Black hams at the Dukeshill Ham Company

will give the ham a lovely golden colour. Do not remove from the smoke house until cold. Remove the muslin cloth.

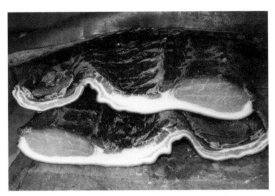

Black bacon showing treacle covering which gives great flavour and also acts as a natural preservative

Black middle ham

Black hams

Gentle Bacon 1

Ingredients

8 gallons of water

22lb Bay salt

6oz saltpetre

2lb molasses sugar

2 pints of treacle

1oz nutmeg

1oz mace

Method

Boil the water, put into a container, and add the Bay salt. Boil the saltpetre as previous instructions and add to the container when cold. Boil the molasses with plenty of water, dissolving thoroughly. Add to the container. Put the treacle in a saucepan with a small amount of water. Heat until dissolved and add to the container, making sure you stir well. Grind finely the mace and nutmeg. Put in a saucepan with a small amount of water. Heat until infused and add to the container. Wait 24 hours for the mixture to cool. Check the density — it should be about 60-70%. This brine will produce a nice gentle bacon. Smaller or larger quantities of this brine can be made providing you keep to the proportions.

MAYNARD'S TIP

Sometimes, new brines do not produce those satisfactory pink colours in the ham. If this happens, take half a gallon of matured brine. Sieve to remove any particles, boil, wait until cold and add to the new brine. Leave for 2 days and your brine will then start to work well.

Brine Cure for Roasting Ham

This will make an excellent ham for roasting, or slicing when cold, and produce an ideal ham for the delicatessen counter.

Ingredients

  10 gallons of water

  20lb rock salt

  6oz saltpetre

  8oz clear honey

  4 pints of dry cider

  20 dried juniper berries

  3oz black peppercorns

  10 bay leaves

Method

Boil the water, put into a container, and add the salt. Grind the saltpetre finely, and dissolve as previously instructed. When cold add to the brine. Stand your jar of honey in hot water to loosen.

Pour it into a saucepan, add a small quantity of water, and warm over a low heat until liquidised; when cold, add this to the brine. Add cider to the brine but do not boil. Grind the juniper berries, peppercorns and bay leaves finely, put into a saucepan with a small amount of water, and boil until infused.

Put into the brine when cold. Stir all the ingredients well and leave for 24 hours. The brine is then ready for use.

If you are going to use this brine for pumping, pump the hams at about 40-50% density. To do this, remove a quantity of brine and add sterile water until the required density is reached.

Curing these hams will take about 5 days. When cured, fetch them out, wash off, and hang to dry and mature. If you want to give them extra flavour you can smoke these, and I found the apple wood smoke gave an excellent flavour to this ham. Smoke for about 2 days; then remove from the smoke house and keep in a cool airy place to mature.

This is an excellent ham for hotels and delicatessens and you will find they will be in great demand. I used to produce a tremendous amount of these hams and they used to give me great satisfaction.

Strong Brine for Smoked Bacon

I used to use this recipe in connection with strongly smoked products, including hams, and it gives them a wonderful flavour. Bacon for smoking is always best prepared in a wet brine, as dry salted bacon tends to come out of the smoke house with a chewy consistency.

Ingredients

40 gallons of water

56lb salt (preferably rock or bay salt)

16oz saltpetre

9lb Muscovado sugar

9lb molasses sugar

½oz pimento

½oz juniper berries

½oz coriander

Method

Boil the water and add the salt. Grind the saltpetre finely, put in a saucepan and heat until dissolved; then, when cold, add to the brine. Put the Muscovado sugar in a large container with some water and heat until dissolved.

When cold, add to the brine. Repeat the process with the molasses sugar, and then add to the brine when cold. Put the juniper berries, coriander and pimento in a grinder, and grind finely.

Put in a saucepan with some water, heat thoroughly until mixed and, when cold, add to the brine. Once all the ingredients have been added to the brine, stir well, making sure everything is thoroughly mixed.

Always stir a brine thoroughly prior to use. It is always better to leave the bone in place as it helps to keep the shape of the ham in the smoke house. If you are going to pump these hams, the density of the brine should be about 40%.

The hams will take about 5 days to cure. After 5 days remove from the brine, wash off and dry. Mature them as usual, making sure they are thoroughly dry before you put them in the smoke house.

If you want the smoked bacon to have a mahogany colour it is a good idea to use pea meal. Put this meal in a shaker and sprinkle it over the middles, on the rind side, before you smoke them. The same goes for the hams: leave for a couple of hours coated with pea flour before you smoke them. When you have finished using it, cover the pea meal shaker with cling film to keep pea meal fresh and dry.

If you do not want a heavy smoke, put your middles in a muslin cloth. Hang them up about a foot apart. If you want a heavier smoke, put them in a smoke net that has a large-gauge mesh. This will allow more smoke to the bacon.

Do not hang by the muslin cloth; always string them up. Also, dip your muslin in cider vinegar — this will ensure

MAYNARD'S TIP

As this brine is mixed specially for smoked products, when removing your bacon from the brine, mark it with an 'S' to denote it is for smoking.

that it does not stick to the bacon.

Once your bacon is smoked, it is better to store it on its own. If that is not possible, bone it out and vacuum pack it. If smoked bacon is next to green (unsmoked) bacon, the smoke will 'jump' — the same applies if there is fresh meat in the fridge. On the hams, put a piece of muslin over the bone. This prevents the bone piercing the vacuum bag.

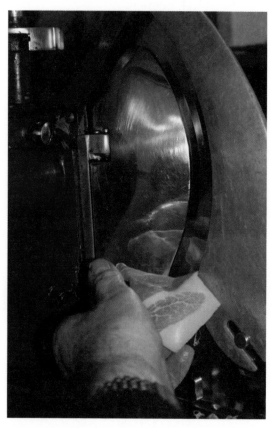

The bacon slicer is the perfect machine for preparing your high-value smoked bacon in fine slices for the consumer

Ancient Barrel Pork

My next recipe is a historical recipe which is no longer needed today as we have other methods of preserving food for long journeys. The recipe is for a barrel pork brine and was used by the Royal Navy, by merchant ships and others taking long voyages across the sea. It was used mainly for pork and sometimes beef.

It was a very successful cure though also a very harsh and salty cure, but it had to be as sometimes the meat was taken to the tropics where the temperatures were very high. There was a very high concentrate of salt in this barrel pork to preserve it. I will give you the recipe though I do not think you will ever need to use it, but for the sake of posterity it's nice to know how it was done and I pass it to you with courtesy.

In the days when barrel pork was common, there were no such things as brineometers. A fresh egg was used (with its shell on) to test the strength of the brine. The water and brining ingredients (except the salt) were mixed in a barrel and a fresh egg placed in it. Then they added the salt by degrees until the egg floated to the surface — once the egg floated they knew the salt content of the brine was right.

To cure the meat, a portion of the brine was placed in the bottom of a barrel. Then the larger pieces of meat were placed in the middle and the smaller pieces round the sides, building them up in the barrel. Once this was done, the brine was poured in and then the lid was put on the top. They sealed the lid of the barrel and then they sealed the veins of the barrel with tar, using what was called a tar spoon and very, very hot tar. They then waited until it cooled and then stored the barrel in the cellars to mature.

Every day a man especially employed to do this task rolled the barrel up and down the cellars; this was to distribute the brine. This was the forerunner of tumbling as it is done today (*see page 52*). The barrels were rolled up and down for a week to a fortnight. That is how barrel pork was done and it lasted for a very long time.

This recipe is part of your heritage and is not recommended for use today, as the salt content is very, very high. The old curers passed this recipe to me and I now pass it on for posterity. There used to be firms in Liverpool and Bristol which specialised in barrel pork. They were very skilled at this and it was a branch of the industry which blossomed for two or three hundred years. I knew a man whose ancestors in Liverpool had a barrel pork company; they supplied most parts of Africa with barrel pork.

Ingredients

8 gallons of water

25lb large grain sea salt, or a combination of salt & water sufficient for an egg to float (*see previous page*)

6oz saltpetre

4oz coriander

4oz basil

4oz bay leaves

8oz black pepper

6oz juniper berries

Method

Soak the pork or beef in a strong solution of salt to cleanse. Put the sea salt into a container. Grind all the herbs and pepper, crushing to a fine consistency. Put them in the container, adding the water and stirring well, and making sure the salt is dissolved. The original barrel pork would have been very salty because of the amount of time in the cure.

London Spiced Bacon

This is a bacon we used to supply to the London markets. It is a delicious, lean bacon with a very distinct taste. It should be cut very thinly.

Ingredients

15 gallons of water

30lb bay salt

6lb light Muscovado sugar

4oz pepper

6oz saltpetre

6oz allspice

4oz coriander

Method

Boil the water and add the Bay salt. Stir well. Dissolve the saltpetre separately and wait until cooled. Add to the brine, stir well and leave for 24 hours to cool. Boil the sugar in a little water, wait until cooled, and add to the brine. The density of the brine should be about 65-70%. Grind the pepper, allspice and coriander; then put in a shaker for sprinkling on the bacon.

Put your lean bacon in the brine. The middles you leave in for 5 days, or you can pump them. If you decide to pump them, pump at the same density (about 65-70%) — you need to pump them at a higher density than you normally would use as the bacon is cut very thinly so more flavour is needed.

Now fetch out of the brine, wash off and hang up to dry. When thoroughly dry, remove rind and surplus fat with a good sharp knife. Take your meat hammer and make the meat level. Take your spices and sprinkle lightly over the inside. Take the thick end of the meat, fold over, and roll up. Start in the middle and start stringing; work along the roll until it is tightly strung. Loop over the last string you put on the roll so that you can use it to hang with. Once this is done, put in a smoking net and hang up to mature.

Mature the bacon for about a week, at a temperature range of 50-60°F. You now have two options: you can leave the bacon green or you can lightly smoke it. I used to smoke using a beech and barley straw and smoke for about a day and a half; then I would leave the bacon in the smoke house to cool off. I was always very happy with this product and it was one of my main lines, but you must try it for yourself. Whichever way you do it, vacuum pack the bacon when it is ready, making sure it is properly marked, and keep in the fridge.

MAYNARD'S TIP

Smoke this bacon with beech with barley straw on top: this produces a distinctive mild taste. Smoke for a couple of days.

Gentle Bacon 2

Ingredients

15 gallons of water

10lb rock salt

10lb bay salt

8oz saltpetre

4 pints of golden syrup

2oz coriander

2oz mace

Method

Boil the water to sterilise it; wait until cooled, then put it in a container. Put the golden syrup in a saucepan and add a small quantity of water. Boil until dissolved and when cold add to the container. Add salt to the container and stir well. Grind coriander and mace finely, put in a saucepan with a small amount of water, and boil until infused. When cold add to the container. Dissolve the saltpetre in boiling water, as previously described, leave to cool and add to the brine.

Test for density: a good brine for this bacon will be about 70%, so keep it at that. If you are using this for pumping, use at 50% as the bacon is a gentle bacon and you don't want to over-salt. The middles will take about 4-5 days, as will the hams. Fetch them out, wash off as usual, and leave to dry and mature. If not wanted immediately, vacuum pack and store in the fridge. Remember to mark them.

Granny's Bacon

This is a nice mild bacon with a very distinctive flavour.

Ingredients

20 gallons of water

20lb rock salt

20lb bay salt

12oz saltpetre

6lb Demerara sugar

4 pints of apple juice

6oz caraway seeds

Method

Boil the water to sterilise it and, when cold, put it in a container. Add the salt and stir well. Dissolve the saltpetre in boiling water. Leave to go cold and add to the container. Boil the sugar in some water until dissolved. When cold, add this to the container and stir well. Add the apple juice to the container and stir thoroughly. Grind the caraway seeds and put in a saucepan. Add a small quantity of water and boil until infused. Leave to cool and add to the brine. The density of the brine should be about 65-70%. This recipe will produce a nice distinctive bacon and is approximately two hundred years old.

Breakfast Bacon Cure

Ingredients

> 20 gallons of water
>
> 42lb rock salt
>
> 12oz saltpetre
>
> 18lb molasses sugar
>
> 2oz mace
>
> 2oz juniper berries

Method

Boil the water, put in a container and leave to cool. Dissolve the saltpetre in your special saucepan, skim any residue off the surface, leave to cool and add to the water — stir well. Boil the molasses sugar in a large container with some water, stirring regularly. This will take a long time and it is better to keep stirring and stay with this, as sometimes the sugar sticks to the bottom of the saucepan. When dissolved, wait until this goes cold and add to the container. Grind the mace and juniper berries and add these to the brine. Wait until the brine is cold before you measure the density. If curing in a tank, the density should be 65-70%. If pumping, the density should be 55%.

Hampshire Sweet Cure

This produces a gentle kind of bacon and is a well-tried recipe from Hampshire. It gives consistently good results and you may consider using it on a regular basis.

Ingredients

> 10 gallons of water
>
> 21lb bay salt
>
> 6oz saltpetre
>
> 3lb black treacle
>
> 1 pint of sweet cider

Method

Boil the water, leave to cool and put into a container. Add the salt and stir thoroughly. Dissolve the saltpetre as previously instructed, leave to cool and add this to the container. Put the treacle in a saucepan with a small amount of water. Heat until dissolved, stirring well, and do not leave to burn. Once dissolved, leave to cool, then add to the container. Now add cider to the container — a sweet cider is best. Do not boil this but put straight into the brine. Leave for 24 hours; then measure the density.

If using this brine as a wet cure, the density should be about 65-70%. If pumping, the density should be 45-50%. Both methods will give you a gentle kind of bacon. You can also smoke this bacon and smoking it will give it a nice aroma and a lovely flavour. I used to smoke this by putting a muslin cloth over the bacon so that the smoke would not be so fierce. The bacon would come out a lovely golden colour.

STOCK CONTROL

This will be of interest to people who want to have their stocks of bacon ready for Christmas; I used to find it was advisable to start preparing for Christmas in November. One of the main things to do is to make sure the bacon is in fresh condition and the next step is to bone the bacon out; there is no point in saving bones. Put the bacon in the fridge the night before, so it is cool. Buy some good quality vacuum bags (a good thick gauge). Pack the middles individually in those and date them — they will keep for 13-17 weeks.

Next, we take the hams. I always found it was better to vacuum pack the hams. Put a piece of muslin over the end of the bone; this stops penetration of the vacuum bag. Put them in the fridge, label and date them. Remember it is essential that all the bacon to be packed is put uncovered into the fridge 24 hours prior to packing and make sure the bacon is dry, so you are not packing damp bacon. Next are the picnic hams. I used to pack these in pairs as the bags are expensive. Put a piece of muslin cloth on the hock; this stops the bone piercing the bag.

The advantage with storing in this way is that the bacons and hams come out with a delightful flavour because they have had a long maturing period. Also, this cuts all the rush and push out of Christmas. A satisfactory temperature for your fridge would be about 35-37°F. Do not freeze the bacon because the fat will come away from the lean, the bacon will go darker, and when it comes out of the freezer it will sweat and the flavour will be destroyed.

Spicy Smoked Bacon

This is a bacon that we used to find was popular with people who wanted a mixed grill. It is a spicy kind of bacon, and we used to produce it as a breakfast bacon. We would use streakys for this recipe. I used to level the streakys up square, take the spare ribs out, and cure these with the bones taken out.

Ingredients

10 gallons of water

20lb rock salt

6oz saltpetre

3oz coriander

6oz clear honey

4oz pimento

Method

Boil the water, leave to cool and add to your container. Add the salt and stir well. Boil the saltpetre, leave to cool and add this to the container. Grind finely the coriander and pimento. Put them in a saucepan with a small amount of water, stir until infused and add to the brine.

Put the honey in a saucepan, add water, and heat until dissolved. Leave to cool and add this to the brine, stirring well.

When the brine is cold, test for density. This should be about 65-70%. There is no need to pump these streakys as they are thin. Leave them in the brine for about 4 days.

After 4 days remove from the tank, wash off and dry. Do not put a hook in the streakys but thread them with a fine cotton twine before hanging up to dry. When dry, grind some peppercorns and spread over them. Now put them in a muslin net and smoke them. Smoke with oak and they will go a lovely mahogany colour, and with the peppercorns and the herbs they will take on a distinctive flavour. We used to call this bacon 'Spicy Smoked Bacon' and it produced a lovely aroma during the cooking.

Spicy Brine 2

This can be used for many things in the food industry, including topsides of beef, silversides of beef, and tongues. It is a very spicy brine and you can also cure fancy hams with it.

Ingredients

40 gallons of water

3lb bay salt

56lb fine salt (I recommend rock salt, finely ground)

1lb 1oz saltpetre

½oz sodium nitrite

8lb dark Muscovado sugar

½oz coriander

½oz pimento

1lb raisins

1lb currants

Method

Boil the water and put in a container. Add the Bay salt and stir thoroughly. Now add the fine salt and stir well. Dissolve the saltpetre as previous instructions, and add to the brine when cooled. Put ½oz of sodium nitrite into a small amount of water and boil until dissolved. Wait until cool then add this to the brine. Dissolve the sugar in a small amount of water, and add to the brine when cool. Grind the pimento and coriander. Put in a saucepan with a small amount of water. Heat until infused and add to the brine.

Now for the raisins and currants. You need to put these into water and leave them to soak, and this is best done the night before — if you put the liquid straight into the brine you do not achieve the full flavour. Boil them the following day. After boiling, leave them to cool and then put them through a sieve. Discard the currants and raisins, but put the liquid into the container and this will give you a nice spicy pickle.

The density for this brine should be 65-70%; this will give you good results. If you are using this as a pumping pickle — e.g. for topsides and silversides of beef or for tongues — I suggest you pump at 45% for the best results. Mark the container well, and when you take the brine out always check the density.

A Quaker Recipe

This was given to me when I worked in America in 1956. It is a general recipe and can be used for a number of things, including breakfast bacon, hams, brisket of beef, and tongue. It originally came from England and the Quaker lady who gave it to me told me that it had originated about 1820. It was in the back of her family's bible and when they went to America it stayed with them. I was very grateful and I have tried it out on a number of occasions.

Ingredients

9 gallons of water

9lb fine salt (I recommend rock or bay salt, finely ground)

3lb molasses sugar

3oz saltpetre

Method

Mix all the ingredients together, as in previous instructions. The density for this brine should be 70%. It is an all-purpose brine and can be used for most things. If you wish to pump with it the density should be 55%.

MAYNARD'S TIP

Do not put raisins and currants directly in the brine. If you do, and decide to use the brine for pumping, you will find they will lodge in the pump and cause plenty of problems. That is why this recipe just uses the liquor from them.

Theo's Favourite Ham

This recipe was my old boss Theo's favourite ham and we produced many thousands of these hams at his bacon curing factory. We sent them mainly to markets in London and the south — they were popular in the taverns, the catering establishments and the high-class hotels. These hams produced an abundance of flavour and in my view they were one of the best hams we produced.

Ingredients

10 gallons of water

20lb rock salt

6oz saltpetre

4lb Muscovado sugar

4oz Jamaican pepper

2oz juniper berries (dried)

4oz coriander (dried)

2oz bay leaves (dried)

2oz black peppercorns

A nice rack of mixed hams maturing. Note the netting on the middle rows which tightly holds together the hams and the sloping trays which take away any fluid

Method

Boil the water, put in a clean container, and add the salt. Dissolve the saltpetre and add to the container when cold. Boil the sugar with some of the water until it is dissolved, then add to the main container and stir well. Grind the herbs, spices and peppercorns, place them in a saucepan with a small quantity of water and boil. Wait until cold and add to the brine. Now wait for 24 hours.

The ideal cut for this recipe is a 16–18lb ham with a short hock. Immerse the pork in the brine for approximately 10 days; you can leave longer if a saltier or spicier product is required. Now remove the ham from the brine, wash with cold water and hang to mature and equalise. Alternatively you can smoke it, first making sure that the ham is thoroughly dry. Smoke the ham using oak chippings for 36 hours; this will give it a lovely mahogany colour. I recommend a 70% density for the brine in this recipe.

MAYNARD'S TIP

The board used to press the bacons down in brine was sycamore (or sometimes beech) as there is little resin in this wood and it does not affect the bacon.

Hams in a large maturing room where the atmosphere must be dry and the temperature around 40°F. Note the nets which keep the shape of the hams and avoid having to use metal hooks

DRY CURED BACON IN THE BAG

I shall finish this chapter by returning to the dry cures. In this section, I shall discuss the art of curing bacon in a bag: the method of using a dry salting mix and a polythene bag. This method of curing was adopted by the big factories as it speeds up the curing process. Try it for yourself – it will give you a very nice bacon.

First obtain your pig from a good source or use one of your home-grown pigs, preferably a gilt — these cure better and you do not run the risk of boar taint. The next step is to divide the pig into pieces. The best cuts for curing in a bag are legs, middles and streakys; the size and shape of the cuts are entirely your own choice.

Ingredients

> 8lb fine salt
>
> 2oz saltpetre
>
> 2lb sugar of your own choice (granulated, molasses, Demerara, light or dark Muscovado or golden caster)

Method

Before starting this recipe, you should cleanse the pork of all debris using a cleansing pickle (*see page 35*). Remember that this process needs to start 24 hours before you plan to cure. Once the pork has been cleansed, remove it from the pickle, lay it on a bench, rind side up, and leave to dry thoroughly (about 1½ hours).

Now put the sugar into a large bowl, grind the saltpetre to a fine consistency, and mix thoroughly with the salt, making sure there is an even distribution of all three ingredients. It is very important to mix well as an uneven mix will not lead to an effective cure.

Each cut of meat you are going to dry salt will require a different amount of the dry salting mixture. You will need 1½oz of dry salting mixture to 1lb of meat. To judge the amount of time the meat needs to be left in the bag, I use the **inch formula** — for every inch of meat thickness, allow 7 days. So for 3 inches, allow 21 days.

If you want a mild flavour then the meat does not need to be left in the mixture for so long. Getting this right is a matter of judgement and experience.

The next step is to take your cuts: the ham, back, streaky and shoulder. First of all, take your ham and apply the mixture, 1½oz per pound of meat (e.g. if the ham weighs 14lbs, use 1lb 5oz of the mixture). Split this amount in half, using one half for the first part of the process. Rub well into the ham making sure the cut surface of the ham is given plenty of mixture. Also, make sure you use plenty and rub well in around the bone areas and the hock.

The next step is to put the ham into a large, good quality polythene bag, making sure the bag is not flimsy. Put a layer of the mixture in the bottom of the bag and then place the ham on the top. Tie up the bag tightly, making sure all the air is excluded — now put this to one side.

Bacon in the bag – be sure to use a tough polythene, one that will remain firmly sealed and will not tear

The next step is to cure the shoulder using exactly the same method and making sure the mixture is well rubbed into the cut surfaces, especially around the blade bone area where the blade bone has been removed. Insert 1½oz of the mixture into this cavity. Take a fresh bag, put some mixture in the bottom of the bag, and follow the same procedure as previously explained.

Next take the streaky. I always remove the bone from this to prevent the bag being torn and I am also very partial to spare ribs! Rub the mixture in well and repeat the above process. Always place

The bacon is well covered with the dry cure and safely sealed in the bag

rind down as the cure is more effective this way.

Now take the back, lay rind down and rub the mixture well in, place in a fresh bag, and proceed as above.

When all the cuts have been prepared, place the bags in the fridge. It is advisable when curing to keep the temperature in the fridge around 40°F to 42°F. If it is above that, it hastens the curing process, and if it is below, curing stops.

After about 4 days you will find that a lot of body fluids have leached out in the bags. Undo the bags and drain off, then remove and re-salt the pieces. Do not overdo this — just a light covering is necessary. Re-tie, and treat the other pieces in the same fashion.

The length of time in the cure is a matter of judgement as I have said before. After 5 days remove the streaky, wash thoroughly in clean water, put a string through one end and hang up to dry. Repeat the process for the other pieces. The temperature should be about 50°F. Hang the pieces in an area where there is a through current of air, and not in direct sunlight. The meat will take about one week to dry off and to equalise (i.e. for the salt and saltpetre to permeate the meat, so that the cure is even and produces a good colour without patches).

The first one to be ready will be the streaky as this is smaller than the other pieces. It is a lovely feeling when you do this for the first time — cut off a slice, and you have just made your first bacon!

After about a week, put 4oz of olive oil into a saucepan. Heat gently with 2oz malt vinegar and 4oz finely ground black pepper. Remove from the heat, letting the mixture infuse, and leave to cool. Apply to the meat using a clean paintbrush, making sure all the surface is well covered. This keeps any flies away and prevents the growth of mould. Repeat this process with the other pieces. When the pieces are dry, wrap each piece in a muslin cloth tightly. Tie with string on the outside. This protects the bacon and it also looks nice when hung up.

The back and the middle should be ready in another week, but the shoulder and the ham will take longer as they are bigger pieces.

You should now be able to produce your own bacon using any of the above methods and there is nothing more pleasant than the first time you make yourself a bacon sandwich with your own bacon — what a pleasure!

Melvyn Ling with some of his dry salted bacon

MAYNARD'S TIP

To maintain the quality of your product, it is a good idea to vacuum pack the meat after it has been cured and dried. Make sure the pack is coded with time and date. It is also a good idea to have a stock book, and to keep a record of how many you have in cure, how many you have maturing and how many you have for sale. This will help you with buying the raw products.

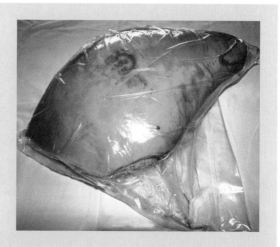

A vacuum-packed ham, the popular way of storing hams and keeping in all the favours and protecting them from outside contamination

PART TWO

TRADITIONAL SAUSAGE-MAKING

Before the ropes can be threaded on to the horn, they must to be taken out of their bundles and have the preserving salt washed off them

Maynard Davies with the essential tools of the sausage-maker: the mincer, bowl chopper and sausage extruder

CHAPTER SIX

SAUSAGES & SALAMIS

Sausage making is one of the most satis-fying things you can do. The first recorded sausage was from Ancient Greece, so the history of the sausage goes back a long way. The Roman soldiers used to keep a smoked sausage in their packs so that after a long day's march they would be able to make themselves a meal. Eaten with wine and seasonings it would last a long time, and I suppose it would have been similar to the salami sausage of today. Closer to home, when the pig was killed the lady of the house used every available piece of meat – some of them even used the squeak!

Equipment for Sausage Production

The basic tools for sausage making are:
- a **mincer**
- a **bowl chopper**
- a **sausage filler**, or **extruder**

First, you will need a good mincer with a number of different-sized plates. These plates will mince the meat in different grades: large, medium and small. After every production, the mincer should be cleaned and never left dirty. You will need a good food oil when you have finished mincing. Apply this to the 'worm' inside which takes the meat up to the grids, and grease it well. It is worthwhile investing in a piece of cotton material to cover the mincer when you have finished cleaning. It will keep dust and debris from the machine, and if you look after your machinery it will last many years.

The next item on the list is a bowl chopper. Many people in the industry do not use a bowl chopper; they use a grinder/mixer. I prefer the bowl chopper

97

as it can do a variety of jobs. It can make sausages, mix all the seasonings, and make pâté. It can also do all of your rind emulsions which I will cover later in this chapter, so I believe it is an ideal piece of machinery to have. It is essential to keep the blades of your bowl chopper very sharp, and I recommend having two sets of blades. Most of the bowl choppers have three blades on the shaft. When you have finished using the bowl chopper, clean thoroughly and then rub liquid paraffin onto the inner bowl so it is nice and clean. Cover over with a clean cotton cloth.

MAYNARD'S TIP

To adjust the blades onto your bowl chopper (below), take a small piece of writing paper. Slip the writing paper inside the bowl chopper and adjust the blades to the paper. This will give you a very fine chop.

Above: The sausage mincer

Right:
The automatic sausage-filler in action

An Essential Precaution

If you are starting a large-scale business from scratch, the sausage department must be in a separate room. This prevents nitrite contamination which could transfer to the sausages from the nearby bacon production. You should also have a bespoke drain in any room where the sausage-making machinery is installed. The drain must be fitted with a fat trap to prevent excess grease from going into the main drain. This must be emptied after each operation and will make daily cleaning easier.

Sausage Filler, or Extruder

The sausage filler is also known as a sausage extruder or sausage stuffer. There are automatic fillers for large-scale production but for medium-scale production I recommend a straight filler where the piston comes up from the bottom. You will need a number of nozzles: large, medium and small. The medium is for all your regular pork sausages and the small nozzle is to make the thinner sausages generally known as chipolatas. The large nozzle is to make salamis and other large products. After use, make sure the nozzles are cleaned thoroughly; apply the food oil and then cover over with a clean cotton cloth.

You will want the sausage stuffer to be able to take the load out of the bowl chopper. This means that if you chop 50lb of meat, the sausage filler should be able to handle the 50lb and keep up with production. If you can only put in 25lb, it will hold you up as you will need to keep refilling. So when you buy your machinery keep this in mind.

When you have these machines installed, have a dead-man's switch on the wall so that when you clean them the power is entirely cut off and there can be no accidents.

Sausage extruder

Spooler holding sausage casings

MAYNARD'S TIP

Use a mobile double-sink when extruding sausage. Extrude into the first sink while a helper links sausage in the second. Put a bucket under each to catch any excess fluid. When finished, the sink can be pushed away. Also, have a water spray in the second sink to clean the sausage of stray pieces of sausage meat and any excess salt. Then hang them up to dry.

Sausage Casings

Traditionally these are natural casings taken from the guts of animals: pigs, sheep and beef cattle. They can be bought as sheep casings, hog casings, or beef casings, and there are different varieties of each. Casings are also known as 'ropes.'

Beef casings are used for black pudding, polony, liver sausage, salami, and some continental sausages. Beef casings come as beef runners, beef middles and beef bungs. Beef runners are used as casings for black pudding; beef bungs, which are larger, are used for luncheon sausage.

For your regular sausage production you can use sheep or hog casings, which also come in different sizes. I always used to use medium-sized hog casings, which gave me 8 average-sized sausages to 1lb of sausage mix, but that was my preference. Sheep casings can be used for beef sausage, pork sausage, and for the smaller sausage generally known as a chipolata. Sheep casings will give you 16 chipolatas to 1lb of sausage mix.

You can also buy synthetic casings which are easy to store and come in all sizes. Keep them in a dry place in the storeroom. In my opinion, they can produce a reasonable product in some cases, but I

MAYNARD'S TIP

When putting casings in the fridge do not stand directly on the floor, stand on a rack, so there is cold air flowing all round the container. Make sure the lid is airtight to prevent smells in the fridge.

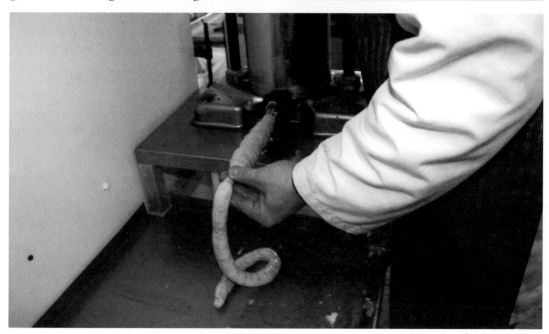

Using the sausage extruder: tighter finger pressure for smaller sausages, looser pressure for the chunkier ones, such as the Cumberlands

Ropes threaded in readiness on to the horn of a sausage extruder

always used the natural casings as I think people like a natural product and I felt they made a far better sausage.

There are some points you must observe when buying sausage casings. It is never wise to buy cheap ones, because they will burst when filling or cooking. You want to buy good quality casings and in the end they will be the best value. They should be white in colour and they should not be strong smelling. If you buy old casings that have been in store for a long time, you will find they will be over-salted and they will break in cooking or when filling on the machines. The best way to buy sausage casings is to buy them to match the level of your production so you never have more than you want. If

you find a good supplier, pay his price; this is not false economy.

To store the casings, I used a plastic dustbin and put them in clockwise so that when you take them out, you take them out anti-clockwise. Try not to put them directly on the bottom – put something on the bottom so if there is any excess water they are not sitting in water. If you need different sizes, have different bins for each one.

The night before your sausage making, put them in a bucket with some sterile water – do not put more than two bundles in each bucket or you will find they will become entangled and will take you ages to sort out.

On production day, fetch the buckets out of the fridge, replenish with clean, sterile water, take one bundle out, and put the other bundle in another bucket. Undo the tab and run the casings out. Put the ends of the casings on the rim of the bucket so they are easy to reach. If you have any casings left over, sieve them and make sure they are dry. Re-salt them and put them on top of your container for re-use.

Casings come in bundles, which I prefer, and you can also buy them on a plastic spool. All you need to do is put them in a bucket of water; but the drawback to this is that they are very expensive so you will have to decide if you can afford them.

Natural sausage casings on the rods

MAYNARD'S TIP

Boil a bucket of hot water, put in the fridge when cold, and run this sterilised water through the casings before use.

How to Make Sausages

The basic ingredients for a good sausage are some lean meat, some fat and some added seasonings and binders in the correct proportions. With careful attention to the proportions, this will give you a delicious and pleasing product to eat.

Poor quality 'waste' products should not be used. You should use nothing but the best meat or you will end up with a sausage with pieces of gristle in it, and we have all had some of that!

The seasonings give the sausage its distinctive flavour. The main ingredients are salt, pepper, dried coriander, mace, nutmeg, paprika, and pimento, but there are many other flavourings that can be added for today's tastes. I will deal here with the traditional flavourings for sausages and include well-established regional sausages.

First, we shall start with the main ingredient and that is the pork. I suggest you select good quality pork. To make 10lb of sausages you will need 7lb of lean meat and 3lb of fat. The lean meat and also the fat must be minced on a large grid; smaller grids may be used for tougher meats.

It is always better to cut up the pork the day before and keep it in the fridge overnight. This is done to keep the meat as cool as possible as it goes through the mincing machine.

If the meat warms as you produce the sausage, the shelf life will be shorter as these sausages contain no artificial preservatives. It is therefore recommended by today's Environmental Health practi-tioners that traditional sausages are eaten within 48-hours of being made, and that they should be kept well refrigerated at all times at 2-4°C. If frozen immediately, however, these sausages can be kept for up to six months.

Put the lean meat through the mincer into an adequate-sized bowl, then apply the seasonings. Use ½oz of seasoning to every 1lb of meat, making sure you mix thoroughly.

Honey and sugar can be added to the mix to sweeten the flavour, but do not put in too much. The honey should be clear – add about ½oz to a 10lb chopping mixture – and the sugar should be granu-lated. Honey and sugar also act as natural preservatives.

The last thing to do is to put the fat in which has been minced. At this stage, you can put all of the lean meat into your bowl chopper, if you have one. I prefer a smooth texture but it is all down to personal preference or regional taste – e.g. the Cumberland sausage is traditionally a coarser sausage and should not be minced too finely.

Once the sausage is finely minced, it is then put into a sausage filler or 'extruder'; this is the machine that puts the sausage meat into the casings, also known as 'ropes' in the trade.

As I have said, it is advisable to have the correct quantity of casings for the amount of meat you have. Casings come in bundles and these have to be soaked

1. Measure a double-sausage length

2. Make a loop

3. Now thread the sausage through

4. Link completed

overnight in water and washed again prior to use.

To put the casings on the sausage filler, use your thumb and first finger and thread them onto the nozzle. Put sausage meat into the filler, making sure it is well pressed down so there is no air trapped in the mixture. If air becomes trapped the skin blows off and you have to start again! Now you are ready to stuff the sausage meat into the skin. Make sure you maintain an even pressure as the filling goes into the skin – a tighter pressure makes a thicker sausage and a looser pressure makes thinner sausages.

The art of linking sausages comes with experience. Learn from someone with experience then practise until you are competent. Modern sausage makers invariably use a filler-linker machine which twists and finishes the sausages automatically.

Some sound advice: have all your ingredients ready – lean meat, fat and seasonings – prior to making sausages. This ensures you produce the product as quickly as possible, keeping the mixture as cool as you can. Any seasonings left over should be stored in an airtight jar, as the flavour soon goes when exposed to the air.

Binders

Binders are added to the meat particularly by large producers and are used to add bulk to the sausage. They also bind the fat and the lean and prevent the sausage casing leaking, but personally I do not use much of them because I believe in a purer product; I used to produce a sausage with 95% meat. The shelf life of this is not so long as one with artificial preservatives but the taste is delicious.

The binders used in sausage production include rusk, rice, milk powder and soya flour. Plain flour can also be used to tighten the mix. If you want to use soya flour or milk powder, apply as a 2% proportion of the whole mixture. The rusk used in sausage production is a straight bread rusk without yeast, as the yeast would make the sausage go off. Rice makes an excellent binder, and if you decide to use it I recommend you adopt the following procedure.

Use 1lb of rice to 3 pints of water. Boil until the rice is ready, then remove from the boil and put onto shallow trays to cool. When cool, put into the fridge. When removing from the fridge, skim any foam from the top of the rice. The high starch content acts as a natural preservative. It is then ready for production. If the rice makes the sausage too white, add a little natural red colour to the mixture. At the time of writing, the only colour you are allowed to use is carmine (an extract of cochineal). This is the 2007 EC regulation.

If you find the sausage meat is too hard when you come to put it in the stuffer, then add 4-6 fl. oz of water. But if you are going to do that, make sure the water has been boiled and cooled in the fridge before adding to the mixture — that will help to soften the meat.

When your sausage is made, it needs to be put in a cool, clean airy place so that the moisture on the skin can dry off – this is important. In modern conditions, and if you have the space, this would be in the fridge and they will dry in there: put them on a flat tray or hang them up on sausage sticks. If you do not have a large fridge, then gently dry the sausage with a clean cloth. Whatever happens, don't leave your sausages out at room temperature. Minced pork does not keep for long.

The sausage we are looking for is a firm sausage, one that does not burst in cooking, that has a good colour and flavour. If you follow the advice in this book, then that is what you will achieve.

Cooking sausage is an art in itself. Some people prick sausages, but if you do that then all the fat drains out and the sausage will be dry. Sausages should be placed on a tray in the oven, not too close together so you can turn them while they are cooking. You can fry sausage but again do not prick the skin as the sausage will cook in its own fat and this is what gives it the taste.

There are many varieties of sausage around, such as tomato, sage, apple, and curried. Every region has its own favourite. There are also many ways of using sausage meat apart from the favourite stuffing for the Christmas turkey. There are many other recipes you can do, including scotch eggs, toad-in-the-hole, sausage rolls, sausage burgers, and sausage plait (which, covered with pastry, makes a very filling meal). Here are some recipes for different kinds of sausage, including a Royal Cambridge sausage, a Cumberland sausage, and many others.

Note that many of these recipes use water in the mixings, and it is common in the industry to measure small amounts of water by their weight in pounds rather than in pints.

1 pint of water = 1¼lbs

1 gallon water = 10lbs

Rind Emulsion

Rind emulsion can be an excellent natural binder. It uses the rind off the pig, including the ears and anything else left over that cannot be used in any other production. There are some golden rules for making this emulsion. Sometimes it does help the end product, and the bonus is that you waste nothing at all. It is an excellent thing to put in black pudding as it does good work binding the blood and the groats together. It is also used in some cooked meats and is an excellent binder for salami and liver sausage.

The recommended proportion is 10% of the whole product, so you can replace 10% of a mixture with rind emulsion. Many manufacturers use rind emulsion for its binding qualities and for the 10% saving on some of their products.

I never added rind emulsion to my normal pork sausages: it is far more appropriate for liver sausage and black pudding. I suggest you check your recipes with your local trading standards officer to ensure compliance with the QUID (Quantitative Ingredient Declarations) meat regulations.

Method

Make sure all the rinds are free from hairs and debris and make sure they are clean. Put the rinds in the boiler, or cauldron, add water and cook for about 40-90 minutes depending on the thickness of the rinds.

When fully cooked, mince on the finest grid you have so that the rinds are cut into very small pieces. Whilst still warm, put them in your bowl chopper, and to 30lb of rind add a certain amount of sausage seasoning of your choice. Turn on the bowl chopper and whilst turning add the seasoning to the rinds and you will find there are little bubbles coming all the way up.

When all the bubbles rise to the surface, that is when the rind emulsion must be taken out and put onto long trays to cool. When laying out the trays, do not put them together but leave a gap as the heat will travel and they will not cool quickly.

I recommend using some rind emulsion as a binder in your liver sausage and black pudding, but not in your conventional sausages.

TRADITIONAL SAUSAGE RECIPES

Classic Pork Sausage

For this simple recipe you will need 14lb of lean pork and 6lb of belly pork. For the seasoning, use the following ingredients:

> 1lb fine salt
>
> 3oz white pepper
>
> 1oz nutmeg
>
> 1oz mace
>
> 3oz coriander

Put all of the seasoning ingredients into a container and mix well. Use ½oz of seasoning to 1lb of sausage meat. Mince the lean meat on a large grid. Do the same with the belly pork. Put the lean meat in your bowl chopper and add the seasoning. When mixed, add the belly pork and grind to the required texture.

MAYNARD'S TIP

Code the seasonings in your seasoning room for customer popularity and this will give you some control on how much to make.

Royal Cambridge Sausage

The following great recipe will produce 60lb of sausage meat. Cooked rice must be kept refrigerated at all times.

Ingredients

> 22lb lean meat
>
> 16lb back fat
>
> 7lb cooked rice
>
> 1lb 14oz seasoning (see recipe in blue box on page 107)
>
> 4oz milk powder
>
> rusk – this is optional, a little can be put in the chopper last

To make the sausage, first take the lean meat and back fat out of the fridge. Keep the lean meat and fat in separate containers. Mince the lean meat on a large grid and do the same with the fat. Put the lean meat in a bowl chopper and give it one revolution. Stop the bowl chopper, lift the lid, take the seasoning and sprinkle it all the way round. Then re-start the bowl chopper and add your rice.

Wait until the rice has bound itself to the lean meat. Put the milk powder in next, and chop until absorbed. Then add your fat but do not chop the fat too finely – keep it so that it is visible. If the mixture is too stiff, add a small amount of sterilised water from the fridge: this will soften the mixture for putting in the extruder. When removing the mixture from the bowl chopper, make sure you scrape the bowl with a spatula to remove all the sausage meat.

Make sure when you pack the sausage filler that there is no air in the mixture, because if you do not expel the air it is a waste of sausage skins and your time. To do this, make small footballs of the sausage meat and pack them tightly into the extruder, expelling all the air.

Once the extruder is full, pat the meat down with your hand. Put the lid on tight, put your foot on the pedal and bring some sausage meat up.

Now take your sausage skin from the bucket, having prepared the skins as described above, and put the skin on the 'horn' – the horn is the pipe on the end of the extruder. Run the skin on the end of the horn and start to extrude the sausage. Try to stuff the sausages out in long rings, a bit like a curled up hosepipe, and then you can link them.

Sometimes when you link them you can get a broken skin. This is not your fault – it happens to all of us. If it happens, I recommend having a hosepipe with a shower head close by so that you can wash the sausage off.

Another good idea is to wash the salt off the sausage casing, and it is also a good idea to have a number of cloths at hand to dry the sausage skin.

When you have linked them, and if it is a good production, I recommend having a sausage stick to hang them on. Do not overfill, and leave space to let the air flow around them.

To make the Seasoning for the Royal Cambridge Sausage

18lb fine salt

7lb white pepper

2oz cayenne pepper

4oz ginger

6oz nutmeg

6oz mace

4oz ground coriander seed

Method

To make this seasoning simply mix the above ingredients together thoroughly and keep in an airtight container, making sure you mark the container well (note that these quantities are for large-scale production). The proportions are: ½oz of seasoning to 1lb of sausage meat.

MAYNARD'S TIP

Put them in the fridge and space them out to dry – do not put them near the fan as this will dry them out completely. Do not leave the fridge light on, as this will discolour them.

Centurion Sausage

This recipe is for a smoked sausage. It is the oldest sausage recipe I have and it is about 2,000 years old. The sausage was produced for the Roman centurions in the Roman army to take on their marches, so at the end of the day they could make themselves a meal. Their rations consisted of a flagon of wine, wheat meal and these sausages.

For this recipe you will need 60lb of pork, and to do it authentically this will have to be cut by hand – the Romans did not have bowl choppers! For seasoning, use the following:

Ingredients

    5lb sea salt

    4oz coriander seeds (ground)

    3oz fennel seeds (ground)

    ½ pint of Italian honey

    1 pint of red wine

    2lb wheat flour

Method

Put the pork in a large bowl, then add the salt and mix by hand until the salt has been absorbed. Spread coriander over the meat, and then do the same with the fennel. Work both of these into the meat by hand. Pour the honey over the mixture and blend well. Add the wheat flour and blend once more. Pour the red wine in until you have a soft mixture, but do not make it too wet.

Once you have completed this put a cloth over the bowl and leave for 4 hours. The reason for this is that the wheat flour will absorb the moisture, the salt will start to cure the product, and the herbs will start to flavour the sausage, so it is important to leave the mixture for 4 hours in a fridge.

For this sausage you will need beef runners for your casings. They need to be 10-12" long. Prepare them by soaking them, making sure they are clean and pliable.

When the mixture has been left for 4 hours, take a beef runner and tie one end with a cotton string. To make this sausage authentically, you will need a funnel with a large outlet. Insert the funnel into the runner, pack the meat into the funnel making sure there are no air pockets, and push the meat into the runner making sure it is packed well, but not too tightly. When the runner is full, pull off the end and tie, including a loop to hang it up with.

To mature, leave the sausages for 2 days in a fridge. When dry, put in the smoke house and smoke for 3-4 days. I suggest you smoke them using an oak smoke with wheat straw until they are hard. When smoked, leave them in the smoke house to cool down, and when they are cold fetch them out and coat with olive oil. Hang them in the drying room to mature – this will take about three weeks to a month. You will then have an authentic Centurian sausage!

Tomato Sausage

Ingredients

12lb lean pork

5lb back fat

4lb rusk (medium grade)

4lb water (many traditional recipes in the curing industry measure water by weight)

2lb tomato puree

Seasoning (see blue box)

Method

Take the rusk and put in a large bowl. Sieve the tomato puree – it is a good idea to do this as sometimes there are pips and lumps in it. Mix the rusk and tomato puree by hand. Mix the water into the rusk and tomato puree, making sure it is blended well. Once this is done, put in the fridge for an hour.

Take the lean pork and put through the mincer on a fine grid. Then take your fat and put through on a large grid. Put the lean pork in the bowl chopper and run for one revolution so it is evenly spread. Stop the chopper, lift the lid and add the seasoning, applying ½oz to 1lb of meat, but do not over-season as you want the tomato puree to flavour this sausage.

Now add the rusk and tomato mix. Blend well until you have a good, even

colour throughout. Add the fat, and chop this until you have a good consistency. Remove from the bowl chopper and put the mixture into a container. Stuff using very thin sheep casings – they will give you 12 average-sized sausages to 1lb of sausage mix. Do not stuff the tomato sausage too tightly into the sheep casings: give them a little room to expand! Store in a fridge at 2-4°C.

MAYNARD'S TIP

With tomato sausage it is advisable to do the production at the end of the day (when you have no more products to put through the bowl chopper) as the puree will taint all the other products.

To make the Seasoning for the Tomato Sausage

8lb sea salt (crushed very finely)

4lb white pepper

2oz nutmeg

2oz mace

2oz ginger

1oz paprika

Prepare the seasoning by mixing the ingredients together thoroughly. Make sure you keep this seasoning separate from other ingredients and label well. It is used in proportion of ½oz seasoning to 1lb of meat.

MAYNARD'S TIP

You have the option to leave out the water and rusk and just use the lean and the fat: the choice is yours.

Cumberland Sausage

This is a very coarse sausage, containing large lumps of meat and fat, and should be minced on a large grid. You have to take particular care when putting the meat into the bowl chopper that you do not chop it too small.

There is a small amount of rusk and water to put in but in some cases you can dispense with that and just use the lean meat and fat. If you do add rusk and water,

remember that it must be a coarse mix. When stuffed, do not link the sausages but make them into a ring of approximately 1½lb of sausage meat. Tie off, wind round a stainless steel sausage stick, and leave to dry in the fridge.

Cumberland sausage: sometimes made in rings, like the black pudding, but more usually made as a continuous ring (as pictured here)

Ingredients

22lb good quality pork – the quality is important as the sausage is made using the large diameter square grid and all the gristle must be removed.

10lb back fat (no inner fat)

5lb coarse-grade rusk

1 pint of sterilised water

seasoning (see blue box below)

To make seasoning for Cumberland Sausage

Mix the following ingredients thoroughly and apply at a proportion of ½oz seasoning to 1lb of meat/fat. Note that these quantities are for large-scale production:

15lb sea salt (finely ground)

6lb white pepper

3oz sage

2oz mace

Method

Mince the pork using a large grid, and then do the same with the back fat. Put the rusk in a tin and mix with water; leave the rusk to absorb the water so it is pliable.

Rusk comes in different grades of coarseness: superfine, pinhead, medium and granular. They all absorb the same amount of water (2 parts water to 1 part rusk) but the speed at which they absorb water varies. Superfine rusk has instant absorption while granular may take 30 minutes.

Put the pork in the bowl chopper, spread out evenly and give it one rotation, then add seasoning on the top.

Start the bowl chopper and add the bound rusk quickly, then add the fat. Chop to the desired consistency. Do not chop too finely as this is a Cumberland sausage: chop it so that the cubes of fat and meat are in large enough pieces.

Remove from the bowl chopper and put into a container. Leave for a couple of hours before stuffing out. The reason for this is that you want the seasoning to permeate all of the meat.

Next pack into your extruder and do the stuffing, remembering to produce them in large rounds of approximately 1 to 1½lb. Finally, hang them on sausage sticks to dry in a fridge.

Mutton Sausage

This is a Welsh recipe and produces a very tasty sausage. I used to sell this type of sausage to hotels for use in a mixed grill.

Ingredients

    20lb lean mutton

    8lb pork fat

    5lb cooked rice

    seasoning (see blue box below)

Method

Mince the mutton using a very fine grid, and mince the fat using a larger grid. Put the mutton in the bowl chopper. Add the seasoning (see blue box below); then the rice; then the pork fat. Chop to the desired consistency, and stuff out in sheep casings. If sold, these sausages would have to be labelled as containing pork, under current government regulations.

Seasoning for Mutton Sausage

The following quantity is for fairly large-scale production.

    16lb fine salt (I recommend rock or bay salt, finely ground)

    8lb white pepper

    8oz ginger

    2oz marjoram

    2oz savoury

Mix together the ingredients for the seasoning, making sure they are distributed well. Use ½oz to 1lb of sausage meat.

Warwick Sausage

Ingredients

    22lb lean pork

    14lb back fat or belly pork

    6lb cooked rice

    7lb water

    seasoning (see blue box below)

Method

Mince the pork through a large grid. Do the same with the fat. Put the pork into your bowl chopper. Add the seasoning (½oz of seasoning to every 1lb of meat/fat combined); then add the rice until absorbed. If you find the mixture too stiff, add a little boiled then cooled water. Once this is mixed and you have a pliable mix, then add the fat until you have a nice colour; the more you grind it, the lighter the colour. Chop to the colour you want, then take it out. Let it rest for a while and stuff out into narrow hog casings.

Seasoning for Warwick Sausage

Thoroughly mix the following ingredients and use ½oz seasoning to 1lb of sausage meat. Note that these quantities are for large-scale production:

    10lb rock salt (finely ground)

    3lb white pepper

    12oz nutmeg

    4oz coriander

    10oz sage

Succulent Sausage Meat

This sausage meat is used mainly for stuffing turkeys and chickens at Christmas time. I also used it at other times of the year for barbecues as it cooked very well and retained its shape. For the seasoning, use a mix of your own choice.

Ingredients

> 30lb belly pork
>
> 2lb flour
>
> 5lb coarse rusk
>
> 7lb water
>
> seasoning (see blue box)

Method

Boil the water to sterilise it and keep it in the fridge. Mince the belly pork on a medium grid, making sure you remove the teats and dispose of them carefully.

Mix the rusk and the water to a nice consistency and then split the pork into two parts, one part of 20lb, and the other part of 10lb. Put the 20lb portion of pork into the bowl chopper, add the seasoning and then add the water and rusk.

Mill for one revolution and then add the 10lb portion of pork. Mill for one revolution and then add the flour to tighten the mix. Mill until mixed well. Fetch out and the mix is ready for stuffing.

If you want to use it for sausages, use medium hog casings. For sausage meat, put in sausage sleeves, leaving a couple of inches unfilled at each end. The ends of the sleeves are then turned in on themselves and the whole filled sleeve is put in a tight-fitting container. (Elastic bands or metal clips are not a good idea for tying off as they have a habit of turning up in the turkey!) Add several filled sleeves to pack the container snuggly and then place it in the fridge for the meat to harden. Consume within 48 hours.

Seasoning for the Succulent Sausage Meat

Mix the following ingredients and use ½oz seasoning to 1lb of sausage meat.

> 1lb fine salt
>
> 3oz white pepper
>
> 1oz nutmeg
>
> 1oz mace
>
> 3oz coriander

For sausage meat buy some plastic sausage sleeves. They come in 1lb and 2lb sleeves. Store in a fridge

Staffordshire Honey Sausage

You will find this recipe produces a sausage with a pleasant taste and good keeping qualities. The honey acts as a natural preservative and gives you a nice flavour and a longer shelf life.

Ingredients

    25lb lean pork

    18lb back fat

    8lb medium-grade rusk

    5lbs of water

    6oz clear honey

    4oz milk powder

    seasoning (see blue box)

Method

Put the ingredients for the seasoning in a container and thoroughly mix (make sure you label the container 'honey sausage'). Use ½oz to 1lb of sausage meat. Mince the lean pork on a large grid, and then do the same with the fat. Put the rusk into a large bowl and add the honey and water – this should have been boiled the day before and left to cool. Once the honey and water have been added to the rusk, leave for approximately 10-15 minutes for the rusk to absorb the mixture. Put the lean meat in the bowl chopper and turn for one revolution. Add the seasoning evenly. Now add the rusk and honey. Turn until evenly mixed, then add the milk powder and last of all the fat. Turn until a nice consistency is achieved and then stuff out into medium hog casings.

Store in a fridge at 2-4°C.

Seasoning for Staffordshire Honey Sausage

These quantities are for commercial production, but if you find this is too much you can of course reduce the ingredients proportionately:

35lb rock salt (finely ground)

14lb white pepper

1lb ground coriander

12oz pimento

12oz nutmeg

2oz sage

2oz cayenne pepper (this should be ground just prior to use)

Prison Sausage

Note that this recipe is for information only and not for use!

This sausage was made for Her Majesty's Prisons in the 1950s. It was a very economical sausage to make and that is being polite as I think it should come with a HEALTH WARNING! Nevertheless, just for the fun of it I thought I would include it to demonstrate to you how badly some of our industries have produced this kind of product in the past.

Ingredients

> 12lb pig's head meat
>
> 4lb lean sow pork
>
> 4lb rind emulsion
>
> 12lb back fat
>
> 5lb coarse-grade rusk
>
> 7lb water
>
> seasoning (see blue box)

Method

Put the head meat through the mincer first; then mince the sow meat (this is used because there is less water in sow meat which makes the binding qualities better).

Now take the rind emulsion (see the recipe on page 121 for this). Put this through a large grid, and then put through the back fat separately, also on the large grid.

Now put the head meat, sow meat and emulsion in the bowl chopper. Turn for one revolution and then add the seasoning (use ½oz per 1lb of meat). Turn until

Seasoning for Prison Sausage

5lb rock salt or Bay salt (finely ground)

1lb white pepper

3oz nutmeg

2oz ginger

2oz mace

1oz cayenne pepper

Use ½oz seasoning per 1lb of meat

thoroughly mixed and then add the coarse rusk, having previously added the water to it. Put the back fat in last, and mix until you reach a reasonable consistency. If the mix is too stiff, add a little water to soften it. Stuff into hog casings.

MAYNARD'S TIP

If you have ever wondered what goes into a cheap sausage, then here is a possible answer.

In other words, be warned: with meat products, you get what you pay for.

Pork & Apple Sausage

This sausage has a pleasant texture and gives a lovely aroma of apple and pork, which always seem to marry well. It is a nice sausage to produce near Christmas time. The recipe is for large quantities, but you can tailor it to your needs.

Ingredients

    50lb lean pork
    21lb back fat or belly pork
    14lb rusk
    15lb water
    2lb milk powder
    1 litre bottle of natural apple juice
    seasoning (see blue box)

Method

Mince the lean pork through a large grid. Mince the back fat through a medium grid. Put the rusk in a large bowl. Add half the apple juice; then take half the amount of water and add this. Mix well until you reach the desired consistency. Put the lean meat in the bowl chopper; then add the seasoning on top.

Now add the rusk and the rest of the apple juice. Add the milk powder and the remainder of the water. Finally, add the back fat and mix thoroughly until the desired texture has been achieved. Rest the mixture; then stuff out into medium hog or sheep casings, whichever you prefer.

As with all these sausages, store in a fridge (2-4°C).

Seasoning for Pork & Apple Sausage

For the seasoning, mix thoroughly the following ingredients and keep in an airtight container – use ½oz to 1lb of sausage meat. Note that these quantities are for large-scale production:

15lb fine salt (I recommend rock or bay salt, finely ground)
12lb white pepper
8oz mace
6oz nutmeg
2oz cayenne pepper

American Sausage 1

This recipe uses maple syrup or maple syrup extract and it produces a pleasant sausage with a lot of flavour. American sausage is ideal as a breakfast sausage. I always feel it pleases more than it disappoints.

Ingredients

> 22lb lean pork
>
> 15lb back fat
>
> 6lb medium-grade rusk
>
> 6lb water
>
> 6oz milk powder
>
> 6oz maple syrup or extract
>
> seasoning (see blue box)

Method

Mince the lean pork on a large grid. Do the same with the fat. Put the rusk in a bowl, add the water (which has been boiled previously and left to cool) and maple syrup, and make sure the water and maple syrup are mixed well. Weigh out your seasoning mixture, then the milk powder.

Put the minced pork in your bowl chopper, turn for one or two revolutions, and add the seasoning. Start the bowl

MAYNARD'S TIP

When you have made up your seasoning, make sure the container is well labelled. If, in a future mixing, you apply the wrong seasoning, you will be in trouble!

Seasoning for American Sausage 1

The following quantity is for large-scale production, it can be reduced proportionately:

> 12lb rock salt
>
> 3lb white pepper
>
> 12oz nutmeg
>
> 5oz coriander
>
> 10oz sage

Use half an ounce of seasoning to every 1lb of meat/fat.

chopper and add the rusk and maple syrup on the top. Then add the milk powder and turn until absorbed. Now add the fat. Do not let the fat disappear but keep it as large as you can. Remove from the bowl and let the mixture rest. Stuff out into sheep casings. I prefer sheep casings for this sausage as you might find that if you use hog runners (which have a bigger diameter) the sausage may be a bit over-powering – this is because of the maple syrup. Put into sheep casings and you will find this a delightful sausage.

Theo's Traditional Sausage

This is a traditional recipe which I used to make when I worked as an apprentice for Theo. It was an excellent sausage as it was so simple: the simple things in life are the best, as they say, and on this occasion I agree. This sausage was the main stay of the business. We made it every day, and truly it was a sausage among sausages. During my time as an apprentice, sometimes I wished that I had a penny for every pound of these sausage I made – I would have needed a lorry to take my earnings away!

Ingredients

    25lb lean pork – we used mostly shoulder pork

    5lb back fat (hard)

    3lb rice – Theo always used a little bit of natural colouring to pink it up a bit

    ½lb milk powder

    seasoning (see blue box)

Method

Mince the shoulder pork on a large grid. Do the same with the back fat, keeping the meat and the fat separate. Put the lean pork in the bowl chopper. Put the seasoning on the top, then add the rice and milk powder (the rice would have been prepared the day before). Now add the fat. Turn on the bowl chopper and mill until you reach the right consistency. Take out of the bowl and rest for a while. Stuff out into medium hog casings.

    As always, store in a fridge (2-4°C).

Seasoning for Theo's Traditional Sausage

These ingredients produce a very basic seasoning but the flavour is spot on, in my opinion. Use ½oz to 1lb of meat. Note that these quantities are for large-scale production:

17lb rock salt – we used to grind it very finely
7oz ground coriander
6oz ground pimento
6oz nutmeg
1oz cayenne pepper
1oz Jamaican ginger

Melvyn Ling with his own version of the traditional sausage

Scottish Slicing Sausage

The Scots have different tastes and have different ways of preparing sausages. This is an interesting recipe for you to try out.

It is produced in a large diameter casing or beef runner and the sausage is then cut thickly on a machine and then fried. It looks like a darker version of a slice of luncheon meat.

It is inexpensive and easy to produce and was very popular for fried breakfasts in Scotland, especially with the dockers and ship-builders when those industries still thrived.

So this recipe is quite unusual and you will enjoy producing it. You may find a good market for your own version of this product.

Ingredients

> 20lb lean beef
>
> 5lb beef or pork fat
>
> 3lb medium-grade rusk
>
> 4lb water
>
> seasoning (see blue box)

Method

Mix thoroughly the ingredients for the seasoning and store in an airtight container; use ½oz to 1lb of meat. Put the meat through a large grid and add the fat as you mince.

Transfer to your bowl chopper. Divide the rusk into two parts. Turn on the bowl chopper. Add the seasoning, then add the water, then the first part of the rusk, and wait until the water is absorbed.

Seasoning for Scottish Slicing Sausage

2lb fine salt (I recommend rock salt or bay, finely ground)

6oz white pepper

2oz nutmeg

1oz ginger

½oz mace

½oz ground cloves

½oz cayenne pepper

Use ½oz of seasoning to 1lb of meat/fat.

Now add the second part of the rusk. The rusk makes a tight, stiff mixture, which is what you are after for slicing. Take the mixture out and put it into your sausage filler and stuff into large casings – either synthetic or large beef runners.

American Sausage 2

This is another simple kind of recipe, but simple recipes seem to work in our industry and it's when we get too fancy that they don't seem to take off. I am sure you will find this one produces a very appetising and a very meaty sausage. It may seem simple but it has a lovely taste and I hope you enjoy it.

Ingredients

> 75lb lean pork
>
> 25lb back fat
>
> seasoning (see blue box)

Method

Mince the pork on a large grid and transfer to the bowl chopper. Mince the fat, also on a large grid. Add the seasoning to the meat in the bowl chopper (use ½oz to 1lb of meat). Turn one revolution to mix, then add the back fat, and mill to a nice texture. Stuff out using medium hog casings.

Store in a fridge (2-4°C).

Seasoning for American Sausage 2

Note that these quantities are for large-scale production:

> 30oz bay salt (finely ground)
>
> 6oz black pepper
>
> 2oz sage
>
> 1½oz Jamaican ginger

Smoked Sausage

This is an easy sausage to make because smoked products are used in the mixture, so further smoking is not necessary.

As for the smoked products, smoke some streaky bacon or the cheeks from the pig's head. Smoke in a muslin cloth so that the smoke is not too heavy. They are then ready to put in your smoked sausage.

The best way to cure these prior to smoking is to cure them in a sweet brine. Cure them for about 4-5 days and let them mature. For best flavour, they should be cured at around 60% salinity.

Ingredients

> 8lb beef
>
> 8lb lean pork
>
> 4lb smoked belly or cheek
>
> 3lb rusk
>
> seasoning (see blue box)

Method

Mince the meat (beef and pork) on a large grid and the smoked belly or cheeks (the fat) on a fine grid. Transfer the meat to the bowl chopper. Put the seasoning on top, then add the rusk, then the fat. Mill until

---

MAYNARD'S TIP

If your sales of Smoked sausage are slow, vacuum pack it and put it in the fridge. Do not put it in uncovered, as the taste of the smoke will jump to other products.

Maynard by some smoked shoulders: the ideal ingredient for the smoked sausage

Smoked Sausage Seasoning

Use ½oz of seasoning to 1lb of meat, but bear in mind that you already have salt in the smoked bellies and cheeks. Note that these quantities are for large-scale production:

10lb fine salt (I recommend rock or bay salt, finely ground)

5lb white pepper

4oz mace

1oz coriander

2oz cayenne pepper

the desired consistency is reached. Stuff out into large hog casings, about 12" long. Wash off in lukewarm water and dry.

If you find the taste of the smoked sausage is not strong enough, take equal amounts of pork and beef off the ingredients and increase the amount of smoked meat. This will strengthen the smoked flavour.

Full-Flavour Smoked Sausage

This sausage is put in the smoke house after you have made it, and you will find the full flavour will receive a good response from people who like smoked products.

Ingredients

    25lb lean pork

    15lb back fat

    6oz milk powder

    seasoning (see blue box)

Method

Mince the pork on a medium grid; mince the fat on a large grid. Put the pork into the bowl chopper; then add the seasoning and the milk powder. Start the bowl chopper and turn until all the ingredients are mixed. If necessary, add a small amount of water to make the mixture pliable.

Now add the back fat and chop until the desired texture is reached. Put this sausage meat onto a tray and put in the fridge; leave overnight at a temperature of 40°F.

The next day, stuff out into large hog casings. Do not link these but make into a ring. Hang the sausage ring from a smoke stick, using the same string that you used for tying off the sausage. Then wash off – this will remove any surplus salt or fat (if this is left it will hinder the smoke).

When fully dried put in the smoke house. Smoke between 115-120°F for about 4-5 hours; this will give you a golden smoke using oak. If you want a stronger taste then leave for a bit longer.

Wait until it has finished smoking, open the door of the smoke house and leave the sausage inside until it goes cold. It is then ready for packing.

Seasoning for the Full-Flavour Smoked Sausage

    5lb fine salt

    6oz sugar

    1lb white pepper

    2oz nutmeg

    1oz ginger

Use ½oz of seasoning to 1lb of meat/fat.

Lamb Sausage

This sausage is ideal for a mixed grill. I found it worked best as a speciality sausage and people were attracted to it who wanted a change.

I suggest that, to start with, you make it in small quantities until you determine the amount of trade you will attract for it. It is easy enough to double the quantities of the ingredients.

I recommend you use pork fat in this recipe as lamb alone will make the sausage very hard.

Seasoning for the Lamb Sausage

2lb salt

1oz ginger

2oz mace

1oz cayenne pepper

½oz white pepper

Use ½oz of seasoning to 1lb of meat/fat.

Ingredients

10lb lamb

4lb pork fat

1lb rice

seasoning (see blue box)

Method

Mince the lamb on a medium grid. Transfer to the bowl chopper and apply the seasoning. Use ½oz of seasoning to 1lb of chopping mixture. Add the rice and chop until you reach the desired consistency. Mince the fat on a medium grid.

Transfer to the mixture in the bowl chopper. Chop until the desired texture is achieved. Stuff out into sheep casings (if you use hog casings the taste will be too strong).

Store in a fridge (2-4°C). If the lamb sausage is to be sold, it should be clearly marked 'lamb with pork sausage' under current legislation.

King's Pork Sausage

This is a straightforward sausage – no water or rusk is added, but in my experience it will not shrink as I used to make a lot of these. You can double the quantities if you require a larger amount.

Ingredients
14lb lean pork
6lb back fat
4oz milk powder
seasoning (see blue box)

Method
Mince the lean pork on a large grid. Transfer to the bowl chopper, then mince the fat on a large grid. Put the seasoning and milk powder on the meat in the bowl chopper. Start the bowl chopper, mix the ingredients, and then add the fat. Chop until the desired consistency is reached. Link 6-8 sausages per pound. Wash off any fat and debris, hang to dry and you will find this a delightful sausage.

Seasoning for the King's Pork Sausage

1lb salt
7oz white pepper
1oz nutmeg
1oz mace
3ozs coriander

Use ½oz of seasoning to 1lb of meat/fat.

Beef Sausage

This recipe will add another string to your bow and another item to your menu. Beef sausage has a stronger taste than pork sausage, and I recommend you use a good quality beef. These sausages will be a very red colour as the beef makes them red, so you do not need to add colouring.

Ingredients
20lb lean beef
12lb fat – you have the option of using beef or pork fat but I should use pork fat
5lb rusk
3lb water (the old curers always weighed their water)
seasoning (see blue box)

Method
Mince the beef on a large grid and transfer to the bowl chopper. Mince the fat, also

Seasoning for the Beef Sausage

1lb salt
3oz white pepper
3oz nutmeg
3oz ginger
3oz mace
3oz cayenne pepper

Use ½oz of seasoning to 1lb of meat/fat.

on a large grid. Put the rusk in a large bowl, add the water and mix well. Put to one side and let the rusk absorb the water. Add the seasoning to the meat in the bowl chopper. Chop until the seasoning has been absorbed, then add the rusk and chop until this has been absorbed. Now add the fat, and chop until you can see a small grade of fat showing. Remove and stuff into medium hog casings. Hang onto drying sticks, making sure you label them well.

NB. If as part of the ingredients, you use pork fat as I have recommended, you would have to label your product 'beef sausage with pork' to comply with current legislation.

Store in a fridge (2-4°C)

.

Seasoning for the Coaching Inn Sausage

Note that these quantities are for large-scale production:

7lb sea salt, ground finely

12oz light sugar (I recommend light Muscovado or Demerara)

6oz ginger

4oz sage

1lb white pepper

1oz cayenne pepper

Use ½oz of seasoning to 1lb of meat/fat.

Coaching Inn Sausage

This is a very old recipe which I believe was made for the coaching inns of England. The sausage was given to the travellers for their breakfast, and the sugar in it would give them fortitude for their journey: travel in those days was a lot harder than it is today. The sausage has a high percentage of herbs, and the sugar gives it a sweet taste and also acts as a preservative. You will find this an unusual sausage and it should be well received by customers looking for something a little bit different.

Ingredients

25lb shoulder pork – make sure you remove all the gristle

17lb back fat (hard)

6lb rice – well boiled and with a soft texture

6oz rice flour

seasoning (see blue box left)

Method

Mix all the seasonings well and keep in an airtight container. Add ½oz to every 1lb of chopping mix. Mince the lean pork on a medium grid and transfer to the bowl chopper. Mince the fat on a large grid. Add the seasoning to the meat in the bowl chopper; then add the rice.

Chop until mixed, and then add the fat and last of all the rice flour (this will tighten the mix). Chop until you reach the desired texture. Stuff out into large hog casings. Wash off and put on sausage sticks to dry in the fridge.

Midland Sausage

I made this sausage for many years and never had any complaints so I pass it on to you for good fortune. It is a well-tried sausage with a distinctive flavour and I hope you try this because I think it will bring you a lot of success.

Ingredients

> 22lb pork
>
> 12lb back fat
>
> 5lb rice
>
> 4oz milk powder
>
> seasoning (see blue box)

Seasoning for the Midland Sausage

Note that these quantities are for large-scale production:

> 10lb fine salt (I recommend rock or bay salt, finely ground)
>
> 2lb white pepper
>
> 10oz nutmeg
>
> 8oz mace
>
> 10oz coriander
>
> 4oz paprika

Use ½oz of seasoning to 1lb of meat/fat.

Method

Mince the lean meat on a large grid and transfer to the bowl chopper. Mince the fat on a large grid. Turn on the bowl chopper; turn one revolution to distribute the meat evenly; then add the seasoning (use ½oz per pound). Chop for one round and stop.

Now add the milk powder and the rice and chop until mixed. Add the fat, and chop until you reach the desired texture.

If the mix is too stiff, add a small amount of water – I would chop so I could still see the fat, and this will give you a nice supple sausage. Stuff this out using medium hog casings, linking 6 sausages to 1lb of mix. Wash off the salt and debris, and put on sausage sticks to dry.

When dry, put them in the fridge. This is an excellent sausage which should bring you a lot of trade.

Store in a fridge (2-4°C).

Cocktail Sausage

These are very small sausages and are popular at Christmas time. They are very highly spiced and hotels use them some-times in a mixed grill. If you decide to produce them, you will have a good return on your labours as they are expensive to buy.

Ingredients

>15lb lean shoulder pork
>6lb belly pork
>3lb rice
>4lb medium-grade rusk
>seasoning (see blue box)

Method

Put the lean shoulder pork through a fine grid, and then put the belly pork through a medium grid – you want to keep the fat larger so it won't disappear when cooking and will retain the moisture in the sausage. Put the lean pork in the bowl chopper.

Add the belly pork and seasoning (using the ratio ½oz to 1lb) and then the cooked rice. Mix this until a smooth consistency has been achieved. Lastly add the rusk, which has had some water added to it. If the mix still feels tight add some more water.

Once mixed, leave to rest for a while as you want the seasoning to take hold of the mix. Use sheep casings for stuffing. Put them on the sausage extruder using a small nozzle and link into lengths of approximately 2-3". Put on sausage sticks and hang in the fridge to dry. These cock-tail sausages have a good spicy flavour and they provide a good return for your labour, because they command a higher price.

Store in a fridge (2-4°C)

Seasoning for the Cocktail Sausage

Note that these quantities are for large-scale production:

>16lb bay salt, finely ground
>8lb white pepper
>8oz coriander
>6oz pimento
>6oz nutmeg
>1oz thyme

Use ½oz of seasoning to 1lb of meat/fat.

Salamis

Salami is not of course of British origin but it has become very popular in England, so I thought I would include it. There are many types of salami throughout the world, but Italian salamis are among the oldest (the word 'salami' deriving from the Italian word 'salare' meaning to make something salty). Italian salamis are often named after the cities in Italy e.g. Milano.

Then there are German salamis, Hungarian, Dutch and Danish salamis, and one or two different types from America (as when the immigrants went to America they took the salami recipes with

them), and now we even have the English salami.

Salami goes back many centuries. It was the first sausage that was made in Greece, and the Romans adapted it for the Centurion Army as an emergency pack. The advantage of salami is that, if you do a large quantity such as a 100lb or 50lb chop, it will last a long time and you always have it in stock. So I think it is an advantage to have these recipes and then you will be well prepared.

The following recipes were given to me by an Italian called Mauro when I worked in America. He was very clever at producing salami and I had the good fortune to be taught by him. Like everything else, you need someone to tell you and he knew what to tell me. I think these two recipes will give you good results: the first is for a plain salami, and the second is for a smoked one.

Ingredients

25lb pork
15lb beef
10lb back fat (hard)
seasoning (see blue box)

Preparation

A fortnight before you make this recipe, take the back fat and cure it in a dry salt mixture. For the cure, use 3lb salt, 2oz saltpetre, 4oz sugar, and ¼oz sodium nitrite. The fat should be very, very hard, so you need to cure it a fortnight before making the salami.

Method

To make the salami, first mince the beef on a very fine grid. Mince the pork on a ¾ plate. Remove the rind from the hard fat you have previously cured, and cut the fat into squares to put in the mix. Put the

Medium salami

beef and pork in the bowl chopper, and add the seasoning (½oz to 1lb). Red wine can also be added to the sausage mix if the texture is too dry.

Put the fat in the bowl chopper, and mix so this retains a coarse texture. Remove and put into trays to mature. Place in the fridge, remembering to cover this mixture with a polythene bag to prevent nitrite jump in the fridge (if left uncovered, the nitrite can contaminate other products in the fridge). The temperature should be about 40°F. Leave it in until you have the desired red colour.

Take out when the colour is right, and leave it to reach room temperature before you stuff it. Put the mix in the extruder, and make sure there are no air pockets in it. Use beef runners approximately 16-18" long (you can also use synthetic casings – the choice is yours). Knot one end and tie a piece of string around it.

Commence stuffing and make sure you do not introduce any air – it must be a tight fill. When stuffed, tie tightly with good cotton string, leaving a loop for hanging. Wash off with lukewarm rather than cold water. Hang them up to dry in a temperature of 45-50°F and leave to mature. This will take 4-6 weeks.

When matured and fully dry there are two options. The first is to coat them with rice flour and water. Make a paste and paint this on – this will protect them and trap the flavour.

The second way is to use a good quality olive oil and dip them in this. Either way will protect them from harmful mould and keep the flavour in.

Seasoning for Salami

1½lb fine salt

1oz saltpetre

2oz sugar

2oz white pepper

2oz whole peppercorns

1oz nutmeg

1oz ginger

½oz garlic – add this to the seasoning when you are ready to use the bowl chopper

Use ½oz of seasoning to 1lb of meat/fat.

Coarse salami

MAYNARD'S TIP

When hanging salamis to mature, make sure there is at least 6" between them to promote good air flow. Check daily to make sure they are hardening.

Smoked Salami

Ingredients

30lb lean pork

10lb lean beef

10lb back fat

seasoning (see blue box)

Method

Mince the beef through a fine grid. Mince the pork through a large grid, and do the same with the fat. Put the beef and pork in the bowl chopper, add the seasoning and the garlic, and put the fat in last. If you find that the mix is too tight, add water or red wine, according to taste – this will loosen the mix. Chop until you reach the texture you want.

Place the mixture on shallow trays, and cover with polythene to prevent nitrite jump in the fridge (if left uncovered, the nitrite can contaminate other products in the fridge). Place trays in the fridge and leave to mature. When you have the red colour you desire, take out and leave at room temperature for a while.

Now place the mixture in the extruder. Cut runners to 16-18" as above. Stuff out, tie end with string, wash off and hang up to dry. Leave in the fridge for 2 days. After 2 days, remove from the fridge and leave to reach room temperature – this should take about half a day. Now put in the smoke house. Hang at least 10" apart.

Start smoking at 90°F and increase to 160°F – this will make sure the salami is well cooked and smoked. Leave in the smoke house to cool down. Remove and store at a temperature of 45-50°F, in a place that has a steady current of cool air. Many of the larger producers will have a maturing room for this final part of the process.

**Seasoning for
Smoked Salami**

1½lb fine salt

¾oz sodium nitrite

5oz light sugar

3oz white pepper

2oz whole peppercorns

1oz garlic

Use ½oz of seasoning to 1lb of meat/fat.

Marcus Themans of Wenlock Edge Farm, Much Wenlock, with two varieties of his salamis

NATURAL PRESERVATIVES

For those meat producers who would rather not use the modern artificial preservatives I recommend using some of the many natural ones
- just like the curers of days past. The following ingredients all have
preserving qualities which enabled producers to extend the shelf-life of
sausages, salamis, hams and bacons. I am very much in favour of using
these excellent ingredients whenever possible for not only do they help
keep the meat well, but they infuse it with taste, colour and quality.

Salt

Red or white wine

Milk powder

Sugar (unrefined)

Honey

Vinegar (white)

Treacle (particularly in sausages)

Corn sugar

Pepper (cayenne or white)

Herbs and spices

Smoke (wood or peat)

PART THREE

TRADITIONAL BACON SMOKING

Alan Ball checks some backs in his smokehouse at Bings Heath Smokery, Astley, Shropshire

CHAPTER SEVEN

SMOKING

Smoking goes back many centuries. In early Roman times, for instance, there was a communal smoke house and people took their products along to smoke. So why do we smoke?

In the days of our ancestors, it was a way to preserve food and to store food for the winter months as in those days there were no refrigerators, and the only method of keeping food was to smoke it or pickle it. One of the old ways to smoke was to dig a long trench, fill it with logs, put sticks across it, hang the product from the sticks, light the logs and cover with leaves. Very primitive, but effective — this was the first smoke house.

We have developed over the centuries different types of smoke houses and different techniques. Fishermen developed a technique of smoking their fish by washing the fish in salt water, drying it and then hanging it in a cave on a trellis. They would light a fire in the middle of the cave and block up the entrance with wood until the fish was smoked; this was one of the earliest forms of smoking.

Nowadays, we smoke products not just to preserve food but also to trap the flavours, and I always found it was a very rewarding job and one of the most enjoyable things I did in the industry.

I had a lot of satisfaction from producing a good smoke and to those of you who are thinking of doing it I would like to take you through the whole process, as a smoke house is a business within a business and you can earn a very good living by just smoking different products.

The menu for smoked products is tremendous. I will try in this chapter to explain how smoking is done and how a smoke house is constructed, and how we deal with the different products.

What can we smoke? We can smoke beef, rounds of beef, silversides of beef, hearts, kidneys, and ox liver. We can smoke most parts of the pig — hams, backs, streakys, shoulders, tongues, and middles. We can smoke mutton, rabbits, veal, and venison.

In fact, you can smoke most things, but I think you will probably be more interested in smoking hams and pork, salamis, sausages and luncheon sausage. Other products include cheese, almonds, apple rings, Brazil nuts, eggs, hazelnuts, mushrooms, onion rings, peanuts and the one I used to smoke a lot of, which was salt.

There are two types of smoking:

• One is a cold smoke, which is a smoke between 90-110°F. This is mainly for bacons and hams.

• The other is a hot smoke, which is a smoke for salamis, luncheon meats, and smoked sausage.

For beginners, I recommend you start with the cold smoke.

MAYNARD'S TIP

To add flavour when smoking, gather some juniper berries, coriander, and nutmeg — or any spice you think will give the bacon a nice flavour — and put them in with the sawdust and they will give you a lovely aroma.

The Traditional Smoke House

When we smoked years ago, the old-fashioned smoke house was a brick building with a brick floor and a concrete roof. For a factory size, it would be 16ft by 20ft with the roof being 8ft high. The door was made of steel and the start of the chimney was 6-12" off the floor; this was to trap the smoke so you got an even smoke.

The bacon rails were reversible so you could put them in any position. They were made of stainless steel, and if you did a smoking of bacon and then hams you could move the rails to suit the product; this was a proper smoke house. It was pure skill to use these smoke houses. There were no clocks or guides; only knowledge and experience.

In those days, we smoked by putting chopped barley straw on the smoke house floor; then on the top we put oak or hard chippings (using a hardwood such as mahogany, rather than softwood such as pine) and shaped this like a snake.

The next thing we did was to have some fat wrapped in some paper. We lit the straw and chippings with this, and we would not shut the door until the smoke started to rise. Once the smoke was on its way, the damper at the back was opened; this was to rid the smoke house of any moisture. With the door and the damper open, all the moisture on the walls went up the chimney. This was important as you would not achieve a good smoke if

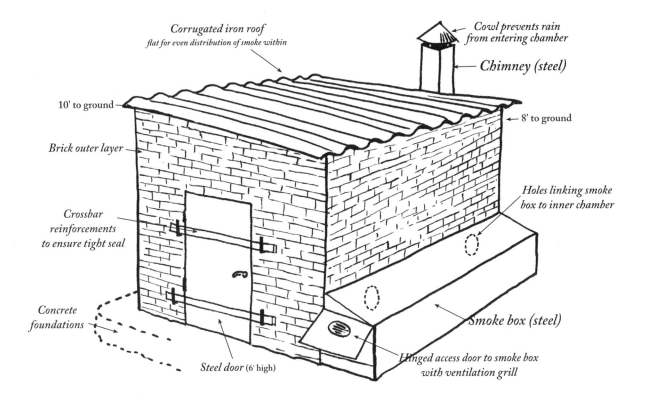

Maynard's perfect smoke house design: ideally it would all be housed under a canopy to protect the smokery from the extremes of weather

the humidity was too high. Basically, the smoke house had to be dry before smoking could commence.

Once the door was shut and the condensation had gone and the smoke was going well, the hams would be put in the smoke house. The hams would be left for a couple of days, and that was how we did it. If we wanted to make a posh job of it, we would coat the hams with very fine pea meal. This would give the hams a lovely deep golden colour, and they would look very nice.

That is how we traditionally smoked.

But over the years I improved the design so we had more control over the smoke. I designed a 'firebox' — a separate compartment for the fuel. This was situated at the side of the smoke house, on the outside, and has many advantages for controlling your smoke, as I will explain.

I will now give you my advice on how to organise your smoke house. If you follow these guidelines, you will find it will be a successful smoke house, which will bring you plenty of happiness and an abundance of rewards.

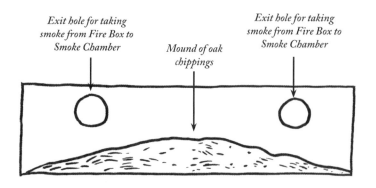

Exit hole for taking smoke from Fire Box to Smoke Chamber

Mound of oak chippings

Exit hole for taking smoke from Fire Box to Smoke Chamber

TopLeft:*Acrosssectionof the fire box, showing the mound of oak chippings and position of the holes which carry the smoke into the main smoke chamber. These can be regulated by using the built-in dampers*

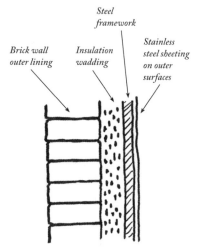

Steel framework

Brick wall outer lining

Insulation wadding

Stainless steel sheeting on outer surfaces

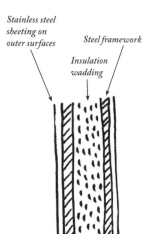

Stainless steel sheeting on outer surfaces

Steel framework

Insulation wadding

Where would I site a traditional smoke house?

If you want to build a smoke house of the traditional kind, you should first consider the location. You will need a site with a good current of air and no obstructions. If you site a smoke house where there is too much wind, the sawdust burns too fast. If the place is too sheltered, the sawdust smoulders.

One method to determine your location is to tie balloons on sticks in different positions and observe the wind direction. Once you have done this, you can site your smoke house so that the 'firebox' gets a constant draught. This is essential, as there are no mechanical parts in a traditional smoke house.

MiddleandBottomLeft:*Crosssectionsofwallingofthesmoke house. The top picture shows the deluxe version, with brick outer walls. Below is the 'economy' version, perhaps a little more inclined to fluctuations of heat, but nevertheless good enough for most production methods*

How to build a smoke house

Throughout my time in the industry I have seen some smoke houses in what I can only describe as foolish places. Most manufacturers, some with only a little knowledge, put smoke houses inside the factory and this is a very foolish practice because we try to keep the factory cool and regulated at a certain temperature — if you put the smoke house inside, it upsets all the plans.

Some manufacturers put the smoke house outside with a door leading from the factory. This is a better solution, but not ideal. In my opinion, the smoke house needs to be away from the manufacturing area.

There are two good reasons not to have a smoke house by the manufacturing area: one is that it creates heat in the factory and this is a disadvantage; secondly, there is a slight risk of fire.

If the smoke house is sited away from the factory this cuts out the fire risk completely. So if we agree that the smoke house should be near enough for ease of operation and far enough away for safety reasons, I think that is a good start.

The perfect position for the smoke house, in my view, is under an open-sided car shelter or shell construction of some kind. This would protect the smoke house from excessive wind, rain, sun and snow and it would keep the temperature of the smoke house nice and even.

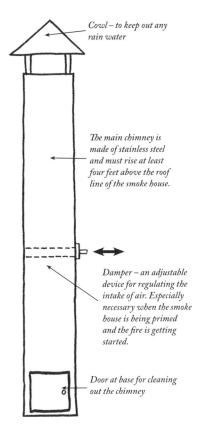

Cowl – to keep out any rain water

The main chimney is made of stainless steel and must rise at least four feet above the roof line of the smoke house.

Damper – an adjustable device for regulating the intake of air. Especially necessary when the smoke house is being primed and the fire is getting started.

Door at base for cleaning out the chimney

The chimney for the smoke house is most cheaply made from thin sheet steel – preferably stainless and fixed to a simple frame of mild steel. The essential elements are marked above

Building materials

I have seen smoke houses made out of wood, concrete blocks, bricks and metal sheets. When I worked in America with the Quakers, they had a lovely smoke house made out of maple logs. It was like a little log cabin, with a well-fitting door and a soil roof, and that is how they made them. It was a lovely smoke house and the

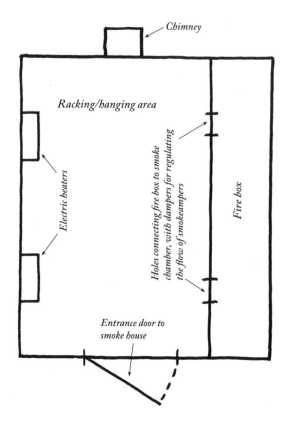

LEFT: *Plan of the smoke house with the fire box running alongside*

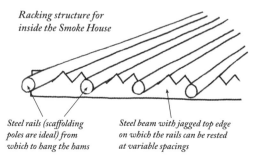

ABOVE: *Diagram of the racking arrangement for the inside of the smoke house*

Size

aroma from the maple seem to permeate the chamber. This was excellent but is not practicable in most cases so we are left with a few choices. I suggest you consider building in brick if it is going to be a long-standing site. Brick gives a better service than other materials. Concrete blocks tend to crack. Steel sheet will give good service, but I prefer brick smoke houses.

The size of the smoke house depends on the size of your production, but if I give you some sizes you can either add or subtract to achieve the desired size as everyone is different. I will work on a smoke house 8ft wide, 12ft long and 8ft high. The reason for the height is that sometimes at peak seasons (e.g. Christmas) you will need to smoke twice your normal quantity. This means you will need space for two rows of hams, one below the other, and if you have the height you can achieve this. In addition, with adequate height you can walk in to load it, and the height seems to produce a better smoke. The door needs to be made of very strong steel so that it is tight when shut and no smoke can escape.

Key elements of the smoke house: the smoke box (above) and the smoke extruder (right).

MAYNARD'S SAFETY TIP

This is a tip the old Curers gave me. I have never experienced it myself, but some of the old Curers told me that if the smoke house ever caught fire, it was very difficult to stop. The old Curers used to have two buckets outside the smoke house door, containing equal amounts of sand and salt. They said that if the sawdust caught fire and you put sand on it, that would extinguish it straightaway. If the hams had caught alight you would throw salt on them rather than ruin them with the sand. You always knew when you visited a smoke house whether the people knew what they were doing. If they did, there would always be two buckets outside the door — one with sand and one with salt — and this was their insurance against a runaway fire.

Design

The smoke house needs a good concrete floor and a good concrete roof. Inside the smoke house you will need adjustable rails, so that the rods can be lifted off to be cleaned. The rails must be substantial as they carry a lot of weight. The walls should be constructed with a double row of bricks and insulation in the cavity.

It is important to put insulation in the smoke house as this holds the temperature. If there is no insulation and you have a cold spot, you will have an uneven smoke. By insulating, you retain the heat and have less condensation. The roof also will need insulating to prevent heat loss.

A smoke house of this size should have two smoke extruders connected to the firebox. The chimney from the firebox will need an access point on the outside so that it can be cleaned out, and I recommend that you also have a clearing out point on the inside.

Inside the smoke house, you will need two electric plugs. I suggest you use waterproof ones; this prevents the smoke penetrating them. They do need to be replaced at regular intervals as they do become corroded.

You will also need two portable electric heaters; these are very important. You put the heaters on approximately 90 minutes prior to smoking. The reason you do this is to prevent condensation; it also helps to warm the smoke house.

The heaters are placed opposite the firebox so we have an even temperature. If you did not do this the heat would be on one side of the smoke house, and the other side would be what is called 'the cold shoulder.' These portable heaters need replacing quite often, so they are more practical than fixed heaters.

You will need dampers on the smoke extruders so that the smoke can be shut off when you have finished smoking; this is important. It is best to put a cowl on the chimney to prevent rain coming down and interfering with the smoking process.

Fuel for the smoke house

The basic material you need for fuel is wood, and the king of the smoke is oak. This comes as shavings or sawdust, and gives your product a lovely flavour. Then there is beech, which is a gentle wood. Sometimes it pays you to mix beech and oak, and this will give you a gentle flavour.

You can also use barley straw: mixed with beech or oak, it gives you a nice combination. Other woods you can use are ash, elm, and sycamore. In America they use hickory and corn stubs, which gives the bacon a pleasant flavour.

For a very hot smoke you can use peat and charcoal. They will give you an internal temperature of 170-190°F. If you want a heavier smoke and more heat, add some logs.

When it is a damp morning and the smoke house is difficult to light, here are some suggestions to ease the pain. The first suggestion is to have some barbecue fluid. Start the fire with this; it is an excellent starter. I used to have a large commercial paper shredder; I shredded all the junk mail, collected it and used this. Another option is to chop some oak sticks and use them for kindling ('morning sticks'). Put them on a tray and pour vegetable oil over them. Leave them to soak and you will find this works very well.

The next little wrinkle for you is to cut small pieces of pork fat, wrap them in paper, and light the shavings this way. So there are plenty of little wrinkles for you to start the smoke house, as there is nothing more annoying when you cannot light the fire.

No smoke without fire

Many people in the industry use what is called a 'liquid smoke.' This is bought in a container — you lay the bacon on a bench and paint it on. The result does look similar to smoked bacon, but it isn't. It is second best, and I never really thought it was the proper thing to do; I thought if you were selling smoked products they should be genuine smoked products so I never used this system.

Some paint liquid smoke on the bacon and put it briefly in the smoke house for a quick flash but that is not what people have paid for. By using liquid smoke, the manufacturer can make his bacon bulkier, increasing his greater yield. But it also makes the bacon a lot wetter and it is then very susceptible to mould so I do not recommend the use of this product.

You can also get a smoked powder; this is used in luncheon meat products, but I do not think it is the right thing to do either. I say, give people what they have paid for.

MAYNARD'S TIP

Store all your wood and sawdust in a very dry place. Damp sawdust burns badly and creates too much moisture in the smoke house.

143

How to Run a Smoke House

In this next part, I will talk you through how to smoke bacon. It is very easy and it is only a matter of common sense. Remember that you have the product in the smoke house and it is a valuable product so we want to make sure we do this correctly.

The first task is to fetch the streakys, hams and middles out of the fridge and leave to reach room temperature. Do not put them in the smoke house straight from the fridge. There are two things you can do to your 'green' bacon before putting it into the smoke house — 'green' bacon is unsmoked bacon. You can apply pea meal or herbs of your own choice. I used to mix mace, coriander and nutmeg in even quantities and rub on the rind, or I would throw juniper berries on the sawdust. Whichever, the choice is yours.

Warming the Smoke House

The next step is to take the bacon to the smoke house. The smoke house I designed has electric heaters. As I explained above, you should turn the heaters on 90 minutes prior to smoking. With the door closed, the smoke house develops a lot of condensation and humidity; the heat from the heaters will clear the walls. You must make sure all the condensation has cleared from the walls and ceiling — if you leave droplets of condensation, they will drop on your bacon and bring some tar onto the bacon and this will mark it.

Fuel

You must now decide which fuel you want to use for the smoke. There are many types, and your choice will depend on the product you want to smoke. Decide if you want to put a layer of barley straw down first and then your sawdust, whatever variety you have chosen.

When I was smoking, I used half beech, half oak and barley straw. The method I used was to put the barley straw in the firebox, about half an inch thick. I used a long rod to level it out; then I added the sawdust. If you use this method, put the sawdust three quarters of the way up the firebox.

Level it out, making sure you leave a space at the front of the firebox to light the fire. Slope the sawdust so it catches once it is lit. To reiterate, the sawdust should be

MAYNARD'S TIP

If a heavier smoke is desired do not put the ham in muslin, but in a smoke net which has a larger mesh, thus achieving a stronger smoke.

144

three quarters of the way up the firebox, sloping down to the end of the firebox so you can light the fire with your own method as described above.

Now go round to the chimney where there is a damper. Open this fully so you have a full blast of air all the way up the chimney; this clears the smoke oven of all condensation.

Once the chamber is clear of any condensation, the next step is to put the bacon in the smoke house. Hang it up at least six inches apart, so that the smoke penetrates all the bacons.

There are two outlets for the smoke. The reason for two smoke extruders is so you achieve an even smoke in the smoke house. The bacon over the smoke extruders will have a heavier smoke than the other products in the smoke house.

MAYNARD'S TIP

Before starting the smoking be sure to remove all dampness from your bacon — do this before you light the fire. The smoke will not take if the bacon is damp.

Maynard in the curing house with a batch of middles ready for the smoke house

To overcome this, wet a piece of muslin cloth with cider vinegar and wrap around your product; tie at the top to secure. But if you prefer a heavier smoke, this precaution is unnecessary. The reason for using cider vinegar is that it prevents the muslin cloth adhering to the ham.

Hanging the bacon

Always use cotton string when smoking, and always leave adequate space when hanging the bacon to allow the smoke to circulate — they should be hung at least 12" from the ceiling and 9-12" apart. If you buy string with plastic in, the heat from the smoke house will melt this and all your products will fall to the floor and you will be in a right pickle.

If the day is damp you may find that the bacon has a film of moisture over it. If so, the best thing to do is to hang it in the smoke house and put the heaters on. This will remove the dampness from the bacon — do this before you light the fire. The smoke will not take if the bacon is damp.

A lot of this is common sense. You will learn by experience how your smoke house operates when you have fired three or four times. You will find when smoking that there is an unevenness to the smoke, and you will find with experience which area smokes heavier and which smokes lighter. I solved this problem by hanging my hams in the centre and the middles either side.

Lighting up

Once you have set out the hams and bacon so that the smoke can filter around each article, the next step is to start the fire. When you light the fire, keep the door on the firebox open and also the main door so you have plenty of draught.

The smoke should now start to drift in. Once this starts to rise, shut the lid of the firebox and shut the door. On the lid is a draught control — open it fully. It is better to start the smoke at a lower temperature (e.g. 80-90°F) and then build up to 100-110°F. Do not allow the smoke house to become too hot. If you do, you will get what is called 'drip' — this happens when the temperature gets so high that the fat begins to melt. The ideal temperature range is 90-110°F. If the smoke is pulled up the chimney too quickly, shut the damper to halfway.

Keep checking

After three or four hours take a peep. Keep having a little peep. This is only for your first days — once you have gained confidence you will not need to do this. For the first few times you smoke, you will find that nosiness will come into it and

MAYNARD'S TIP

Do not put smoked bacon in the fridge because the smoke will leap to other products. Also, I advise you not to vacuum pack smoked bacon, as the bacon will sweat and lose its flavour.

you will want to know how your product is progressing.

Once you are satisfied that all is well, it will take a day or a day and a half to complete the smoke, depending on how your smoke is. This will vary according to the weather, and smoking on a damp or foggy day will take you longer.

Cooling down

When the smoke is finished, open the door, shut all the vents on the firebox, open the damper on the chimney, and wait until the bacon goes cold — never move the bacon when warm. When the bacon is cold, put on storage trucks and move to storage.

Finally, here is another tip for beginners. When smoking and the sawdust is damp you will find the meat will be dark. This is unsuitable — you want a light mahogany colour which is pleasing to the eye. Make sure the sawdust is dry, and store all your wood and sawdust in a dry place.

Maintenance of the smoke house

Once you have built the smoke house there is no money to spend on it; you have had the biggest expense with the building of it. There is very little maintenance to the smoke house, but one of the things you must do is to make sure all the flues are clear. Sometimes a build-up of soot in the chimney can occur. You must always make sure that the chimney has what is

called a 'free flow' — your smoke fans must also have a free flow.

Also, over the years there will be a build-up of sawdust and ash on the walls. When we had done a lot of smoking, I regularly brushed the walls down with a hard brush and made use of the ash on the compost heap.

The removable rails and the smoke sticks need fetching out and steam cleaning, if you have this facility. If you cannot remove the smoke, some caustic soda and a gallon of water will do the job, but make sure the rails and smoke sticks are rinsed well. If after a year you think the smoke house needs a good clean, it is a good idea to use a steam pressure cleaner.

When you have finished using the smoke house, always leave the door open. If you shut the door condensation will form, and this can cause you problems. As you progress with your smoking you will find out how your smoke house works and where the hot spots are and any other foibles she has. You will find that the more you do the better you will be; it is just a matter of common sense.

Other smoked products

You may be called on to smoke products other than bacon; this includes cheese. I had a very profitable business smoking cheese for other people; I also smoked salt and garlic.

To smoke cheese, you will need a table approximately 30" high which should have a stainless steel mesh on the top. You remove the wrappings from the cheese and place it on the table. If smoking more than one cheese, make sure they are 12-15" apart. Smoke the cheese at 80°F for about 6 hours and for most cheeses that should be sufficient.

Remember the golden rule when you have finished smoking, which is — shut the smoke down. Leave the cheese inside the smoke house until it becomes cold; preferably overnight. This makes the cheeses stable, because if you move them before they are cold they can break in half.

I used to smoke a lot of salt for the catering industry. I did this by putting the salt in a shallow dish, covering it with aluminium foil and piercing small holes in the foil. This allows the smoke to enter. Salt would need about 6 hours at the most to smoke. Besides bacon, cheese and salt were the two articles we used to smoke the most, but you can smoke many other things.

Don't penny pinch

Throughout my life in the industry I have come across people who want to penny pinch to produce a smoke house. They adapt old fridges, old toilets at the bottom of the garden, and old barrels turned upside down. My comment on this is that they never ever achieve a good smoke.

To my way of thinking, you need a proper smoke house; then you end up with a proper product and a proper reputation. There are difficulties in converting old fridges: the insulation does not maintain the heat, and then you get what is called a 'shallow smoke' — just on the top of the bacon. Converting a building has another drawback: it does not work

as you do not have the height, the insulation or the correct position, and is never a success. There is also the upside-down barrel and the dustbin lid. Those are for the amateurs; this book is for the people who want to produce good food.

To the people who want to cut corners and build silly structures, all I can say is 'Good Fortune' because you will need it.

PART FOUR

TRADITIONAL
HAMS

CHAPTER EIGHT

COOKED HAMS

In this chapter I will tell you about boiling and roasting hams. This is a very profitable business to be in and it should be approached correctly. It is not just a matter of cutting the ham from the side, putting it in a press and boiling it; it is more complicated than that.

Flavour is a key quality in a roasted or boiled ham, and the other things we look for are good slicing qualities, an attractive appearance, and a long shelf life. There are several ways to cook ham. The traditional method is to roast it; you can also bake it in pastry. You can boil it in a boiler, and you can put it in a press to keep its shape. Hams can also be smoked.

Producing ham correctly will give you great rewards and I will go through each stage. The first thing to do is to select your leg of pork. They need to be about 15-18lb in weight and they want cutting short; this means you cut where the end of the femur bone is. They need to be lean and a good colour. They must be fresh and not 'forward' (this means old pork whose shelf life has more or less run out!). The hams will need soaking in a cleansing pickle for a couple of hours prior to use. So make up a cleansing pickle as described on page 51 and, after soaking, fetch them out.

The next thing to do is to make a sweet pickle for the curing, because the thing you need in these cooked hams is flavour, which will make the product stand out. I recommend the sweet pickle on the following page.

The density of this sweet brine should be 65-70%. For pumping, the density should be about 55%; this will give the hams a delicate flavour. So pump the hams first with the 55% brine; then put them in the denser brine for 6 days.

Take them out and, if you have a tumbler, tumble the hams; this will give them a unified colour. If this is not possible, make sure you pump the hams through the artery prior to insertion in the brine tank. After tumbling, washing off and drying, vacuum pack them and put them in the fridge for approximately 14 days to mature.

The whole object of this maturing time is to make a good colour; you need to leave them in the vacuum bag to mature so that the nitrate has time to work through the meat and produce a good colour. This product will sell on its appearance so the colour is vital.

Sweet Brine Pickle

I recommend this sweet pickle for curing a leg of pork. It will give a pleasant flavour to your cooked ham. If you use the darker Muscovado sugar, it will make the ham darker. Use the following:

10 gallons of water

22lb salt

7oz saltpetre

6lb light sugar (light Muscovado or Demerara)

½oz coriander

½oz pimento

¼oz juniper berries

The Traditional Method of Roasting Ham

In this section, I will tell you how we used to roast hams when I was an apprentice. I am passing this method on as it is part of our heritage and I do not want it to be forgotten. This method I feel is excellent for retaining the flavour of the ham, but you must decide this for yourself.

The hams we baked were between 16lb and 18lb. The first thing we did was to take the ham out of the brine and wash it off, removing any debris it had accumulated while in the brine tank.

The next thing we did was to bone out the ham. We boned the femur out by splitting the hock; then we tunnelled the femur out using a curer's chisel (see photo). Once the bone was out, we tied a piece of cotton string around the end and put the bone back in the ham. The reason for doing this is that the ham will cook better with the bone in as the heat will bounce from the bone; also, replacing the bone means that the ham will keep its shape.

Where the hock has been removed, you sew this up with a stainless steel needle. A lot of manufacturers bone out the hock and replace it in the ham, but I did not do this. We used to supply all the top hotels and restaurants, and when the chefs put the hams on the trays, they had a fine appearance as they had kept their shape.

Once you have boned out the ham and replaced the bone, the next step is to put pieces of cotton string around the ham to hold it or to wrap a cooking net over it — either method will keep its shape. Lay the ham on a shallow dish and put a small amount of salt on the bottom of the pan; this prevents the ham sticking. Add two cups of water to the pan; then place in the oven. These hams take approximately 25 minutes per pound at a temperature of 340–350°F.

When the cooking time is nearly complete (as a general guideline, do this 20 minutes before the ham has finished cooking), remove the ham from the oven, and run the liquid from the pan into a basin, skimming off the fat. Remove the ham from the container and remove the rind; then use the liquid from the ham for basting. To do this, first use a fine sieve to extract any surplus fat.

Now add the ingredients of a basting recipe of your own choice (see below) to the juices from the ham. Mix well and then baste the ham with the basting mixture.

Replace the ham in the oven, cook for a further twenty minutes or so until the ham is a golden brown, and then take out of the oven — this process is called **flashing off**. Remove from the pan at once, place the ham on a clean plate, and wait for the ham to cool.

When the ham is cold, there are two things you can do. You can cover with breadcrumbs of different colours, or you can add cloves and make a diamond-shape pattern on the top. When thoroughly cold, wrap it in greaseproof paper, date it, and then vacuum pack it in the greaseproof paper. Store it in the fridge at a temperature of 35–40°F.

Remember that the bone is still in this ham. When you come to slice it, all you have to do is pull the piece of string attached to the bone — the bone slides out and the ham has kept its shape. This makes a very presentable ham and that is how I did it when I was an apprentice.

This particular roasted ham has been vacuum-packed

BASTING RECIPES FOR ROASTING HAM

Honey Basting Recipe

Ingredients

> 4oz molasses sugar
>
> 4 fl oz orange juice
>
> 4 fl oz water
>
> 3oz honey

Method

Put all the ingredients in a saucepan. Heat until the sugar is dissolved. Put your ham in a container with two cups of water. Place in the oven and cook for 30 minutes at 340–350°F.

Remove and pour the basting mixture over the ham. Return to the oven and cook for 20 minutes per pound. Now remove, rest for 30 minutes and put on a clean plate. Put two very sharp knives in a jug of hot water. Use these to remove skin and excess fat. Dress the ham using breadcrumbs or a glaze.

MAYNARD'S TIP

When removing your ham from the oven, never leave it sitting in its juices — always put it on a clean platter. If you don't do this, the ham will soak up the juices and go soggy.

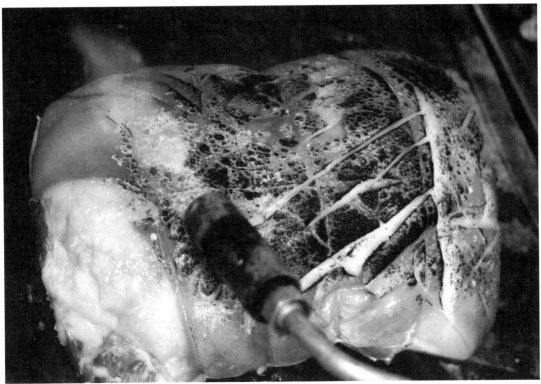

Some manufacturers like to finish the product with a blow torch, giving it a better appearance. Personally, I never saw the need for this and never did it myself

Cloves Basting Recipe

Ingredients

2oz whole cloves

3oz dark sugar

1oz ginger

Method

Remove ham from the oven. Remove rind. Take a sharp knife and make a diamond pattern all over the skin. Put one clove into each of the diamonds and cook in the oven. Mix the rest of the ingredients with the juices from the ham as it is nearly fully cooked. Baste the ham with the mixture, and return to the oven to flash off. Remove when golden brown. Leave to cool; then wrap in greaseproof paper, put in vacuum pack, mark with the date and return to fridge.

Treacle Basting Recipe

Ingredients

    2oz Demerara sugar
    2oz treacle
    4oz golden syrup

Method

Mix the above with the juices from the ham. Put into a saucepan and heat until dissolved. Mix well, baste the ham, and apply the breadcrumbs. Return ham to the oven to flash off. Fetch out, leave to go cold, wrap in greaseproof paper, vacuum pack, date and put in fridge. This is one of my favourite recipes and gives the ham a lovely taste.

Raisin Basting Recipe

Ingredients

    6–8oz seedless raisins
    2 fl oz lemon juice
    3oz clear honey
    ¼oz black pepper
    ½oz coriander

Method

Put the raisins in a bowl with some water and soak overnight. Grind the black pepper and coriander. Remove raisins from the bowl and sieve, keeping the fluid. Add the lemon juice, honey, coriander and black pepper to the fluid from the raisins. Put in a saucepan. Add the juices from the ham, and mix all ingredients thoroughly. Baste the ham with this mixture and return to the oven to flash off. Label and pack as described above. You will find this a delightful recipe.

More of Maynard's Basting Recipes

Basting recipe 1
Juice from two oranges
¼ pint of pineapple juice
¼lb Muscovado sugar
Equal parts of water to the pineapple and orange juice

Basting recipe 2
½ pint of cider
½ pint of pineapple juice
4oz Muscovado sugar

Basting recipe 3
juice of one lemon
juice of one orange
¼lb sugar
Equal parts of water to orange and lemon juice

Basting recipe 4
4oz molasses sugar
½ pint of sweet cider
½ pint of water
Breadcrumbs

Basting recipe 5
½ pint of pure orange juice
½ pint of pineapple juice
4oz molasses sugar

Basting recipe 6
½ pint of pure orange juice
½ pint of lemon juice
4oz dark sugar

Basting recipe 7
4oz clear honey
½ pint of pineapple juice
½ pint of wine vinegar
4oz breadcrumbs

MEMORIES...

Whilst writing these recipes I wandered down memory lane and I remembered a couple, a man and wife. They had a fabulous business roasting and boiling hams and making pâté.

They produced about ten different types of boiled ham, four different types of roast ham and four different types of pâté. It was a very simple operation and their workspace was like a large kitchen. They had three open gas boilers. It was a small operation and the overheads were low but their enthusiasm was high.

I came across them when they wanted me to smoke their hams. The outlets they sold to were hotels, delicatessens, and butchers' shops.

They had one appealing twist to the business — if you bought a ham from them for slicing they provided you with a ham stand, free of charge. I often remember them with affection.

Ham Baked in a Paste

This is a very old method of cooking a ham; we used to produce this very occasionally for people who wanted to carve on the bone. Cooking in a paste gave the ham a very distinctive flavour and it was cooked and carved on the bone and not basted. The ham needs to be large as you need a full slice when carving. The size needs to be about 17–18lb.

We used to lard these hams prior to cooking, using a larding needle. As I explained in Chapter 6, lardings are thin strips of fat bacon, which are threaded through the ham to prevent the ham from becoming too dry. Larding the ham keeps it supple and gives it a pleasant taste. Some of the old Curers used to make the larding bacon with coriander and mace. Some people also rub honey over the ham, prior to cooking, to sweeten the flavour.

Method

Soak the hams for 24 hours prior to use. Wash the cured ham off with a good brine to remove any debris that has accumulated on the surface. Lard the ham and rub some honey over it. Mix flour and water to make a paste, roll out and cover the ham completely. At the top of the ham, make a small chimney with some tin foil, as this will allow some of the stream to escape, thus stopping the pastry from becoming soggy.

Put a small amount of salt on the bottom of your baking tin; this prevents the ham from sticking. Place your ham in the tin and bake for 20–25 minutes per pound (if a large ham, work on 25–35 minutes per pound). Bake at 340–350°F for the prescribed time. Remove from the oven, remove from the pan and place on a clean tray. Remove the pastry cover and rind, and leave to go cold. The ham must be thoroughly cold before you wrap it and put it in the fridge; if you do not do this, the ham will go sour.

This ham is for putting on a stand and cutting with a knife. Do not baste this; the lardings will have added the essential flavour. If you do not lard this ham, you will find it will be very dry. Make sure you wrap the ham in greaseproof paper and vacuum pack it, and that is how you produce a ham for slicing on the bone.

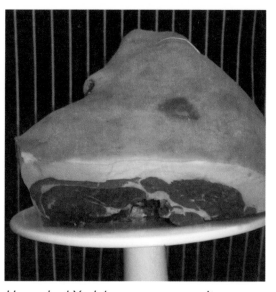

Uncooked York ham - see opposite page

156

Cooking a York Ham

These are cooked quite differently; I have included this method so that if the occasion ever arises you will have the knowledge to produce a York ham. York hams are cured for a very long time (about three months) to produce a distinctive mild, salty flavour. This must be retained in the ham, so it is advisable to follow these instructions.

The first thing to do with a York ham is to remove the muslin that the cured ham is wrapped in. Make a salt brine and wash the ham thoroughly, making sure the white bloom is removed — the white bloom is a harmless mould but must be removed. Once this is done you have a sterilised product. Dry the ham and then put into clear water for 48 hours; some people in the trade add a small quantity of honey. Leave for the first 24 hours; then pour the water away and replenish with fresh water and honey and leave for another 24 hours.

Now remove and dry thoroughly. Spread warmed honey over the ham. If you put a York ham into a boiler with a tremendous amount of water, you lose a good deal of the flavour. So this is the method I used — many people may disagree with it but I found this was the better way. I used to rub the ham with clear honey and place it in a cooking bag, and then dip it very quickly into a bath of very hot water, approximately 200°F. The bag would shrink around the ham and expel all the air. I would then tie the ham with a piece of cotton string, tying it all the way along its length so it would keep its shape. Then I would put it into the boiler and increase the temperature to 210°F.

When it reached 210°F, I would reduce the temperature to 185°F and cook for 20 minutes per pound. Then, when the ham was completely cooked, I would turn off the heat and leave for 30 minutes. After 30 minutes, I would remove the ham and cool it using cold water. Then I would leave it to cool completely. When thoroughly cold, I would leave it in the fridge overnight. The following morning I would take the ham out of the fridge and remove it from the bag.

To carve, I suggest you put your knives in a bucket of very hot water, making sure the knives are very sharp. Remove the skin from the ham — the reason for putting the knives in a bucket of hot water is that the heated knives slide through the rind and take away some of the surplus fat. Now vacuum pack the ham and return it to the fridge. Vacuum packing has several advantages: the ham will mature quicker and retain its flavour, the colour will stay bright, and the ham will not oxidise.

If the York ham was going to be displayed, we would cover it with breadcrumbs, but I never did this until it was ready for sale. I think you will find this is an excellent way of preparing a York ham.

Production of Cooked Hams

Boiled Hams

For this side of the ham boiling business you need an item called a press: by this, I mean a metal or aluminium container that you put the ham in and clamp down. The lid has springs on it and once tightened the container is put in the boiler. The advantage of this method is that the ham retains its shape, retains its flavour and is the right shape to slice. It is an economical way to produce ham for putting on a slicing machine.

Pear-shaped ham press

Method

Soak your ham for 24–48 hours according to size. Remove, dry with a clean towel and bone out. My method of boning out is not the only way to do it, but you must use whichever method suits your production. Take off the hock and keep it for use in other products. Take a very thin sharp knife and a curer's chisel (this is a special chisel for use in curing, see page 48). Remove the femur bone, taking care not to cut the ham inside too much — if you do, you will find that the meat will fall apart when cooked. Loosen the meat by using the chisel around the femur bone. If you do this carefully, the bone will come out easily. The chisel will run along the bone; use your knife to loosen the bone

and it will slide out. When you have completed this task, take a stainless steel needle and sew the end where the hock has come from — now you are ready.

There are two ways of boiling a pressed ham. I used to wrap the ham in a cooking net; then place it in a cooking bag; then put the bag into boiling water. The bag will shrink around the ham, but make sure all the air is removed from the bag. Now put this in the press. Ham presses come in many different shapes, including a pear-shaped one, which is curved at the front, and a square-shaped one. I always used the pear-shaped one (see picture on page 176).

Put the ham in the press rind down. Take the lid, which has two springs on it. Adjust the springs very tight to press down on the ham. Leave for half an hour; this is very important as the ham will take on the shape of the press. Release the pressure a little, as the ham inside the press will expand as it starts to cook.

Put the hams in your boiler. If it is a large boiler, you will have four hams in the bottom. Build them up to halfway in the boiler. Fill with cold water and start the boiler. Now fill a bucket with cold water and stand it on the lid — this will keep the lid down and prevent the steam from

Ham carving knife

MAYNARD'S TIP

Ham presses must always be sterilised before use by placing them in vigorously boiling water for several minutes. Always remember the golden rule of cleanliness.

escaping. Raise the temperature to 200°F. Once this has been reached, leave them like this for 15 minutes; then turn off the boiler. Do not take the hams out, but leave them in the boiler. It is best to cook the hams at the end of the day and leave them overnight, so they do not interfere with production in the day.

In the morning remove the hams and tighten down the lids on the presses. Turn the presses over so that any juices can run out. Then get the hosepipe and spray the presses with cold water. This will take the heat out. It will also seal the hams and trap the flavour, so this is important. When the hams are thoroughly cold, put them in the fridge — still in the presses — for 48 hours.

Remove from the fridge, and dip the press into a boiler that has already reached a temperature of 200°F. Remove the lid and tip the ham out onto a wooden or plastic block. Put your sharp knives in a bucket of hot water; remove the rind and any surplus fat from the ham. Wrap the ham in greaseproof paper; then put a net around the ham — the net will keep the ham's shape.

The hams can be decorated for sale by using breadcrumbs or cloves.

MAYNARD'S TIP

Save the rind off the cheeks and the belly, and when you are cooking a ham put the smoked rinds in the bottom of the press, then put the ham on the top. This will give the ham a nice delicate flavour, and you will have utilised every piece of this product.

Vacuum packing

If you want to slice these hams and pack them yourself, it is better to leave off the crumbs. If you decide to vacuum pack them, then the best thing to use is a gravity feed slicer, always making sure it is kept scrupulously clean and only used for cooked meats.

If you have only one vacuum packer, it must be cleaned thoroughly prior to packing cooked meats. I cannot emphasise enough the necessity for thorough cleanliness when dealing with cooked meats because you can run into many difficulties with bacteria and hygiene if you do not keep up a good standard.

Other Methods of Boiling Hams

I have described my way of boiling hams but other methods are used in the trade. I will now tell you about the most common method. For this, put the hams in the boiler as previously explained. Bring the temperature up to 212°F. After half an hour, reduce it to 180°F (check this with a thermometer). Cooking time should be roughly 15-20 minutes per pound at this temperature; a 16lb ham should take about four hours.

So those are two methods of boiling hams; there is no written law about this, and the choice is yours.

Smoked Boiled Hams

The procedure for smoked boiled ham is the same, but you should have a special press for them. Mark them so that you know which to use for the smoked ham. If you did not want a full smoked ham, you can put a plain ham in the bag and then add a smoked piece of rind on the top — this will give it a slightly smoked flavour.

You can also cook shoulders and middles using the above methods, but if you cook a middle of bacon, it is better to put it in a square press, as the curved press does not do it justice.

MAYNARD'S TIP

Invest in a meat thermometer for your boiler so that you are always sure of the right temperature. Ensure the ham has reached 75°C before it leaves the cooker.

PART FIVE

SPECIALITY
PRODUCTS

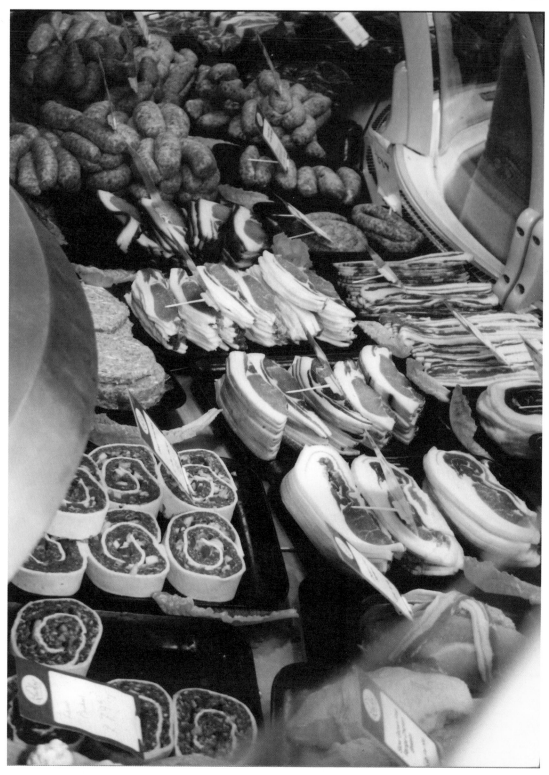

A fine spread of meat products, well-presented for customers at the Ludlow Food Centre

CHAPTER NINE

COOKED MEATS & SPECIALITY PRODUCTS

ROASTING PORK

You will find this is a very profitable article to do. Choose a large leg of pork because you will have the flavour from an older pig and when carving you will have a large slice of pork. Do not try to roast a leg of pork off a young pig as the flavour is not there. The weight you should be aiming for is about 16–18lb.

The first thing to do is to remove the trotter; then you take off the hock, inspect the leg of pork for hairs, and remove these with a sharp knife. If you have many legs to roast, invest in a blowtorch to remove the hairs. It is important to remove the hairs, as there is nothing more off-putting than finding a bristle in your dinner.

Take the bone out of the leg; you will need a sharp knife and a curved chisel — a curer's chisel. Remove the femur bone very carefully, as described in the preparation of hams.

When you have completed this, make a sage & onion stuffing and put it in the cavity. A lot of manufacturers put a little bit of salt in the cavity before putting the stuffing mixture in; this gives a little extra flavour. Many people take the skin off the hock and put it in the cavity, but I am not a big believer in that as I think you disappoint more than you please.

Once the stuffing is in the cavity, use a stainless steel needle to sew the end where the hock was taken off. Tie string around the leg to keep the shape. Using a Stanley knife, score the rind all the way round; if you are feeling artistic you can make a diamond shape.

The next step is to paint olive oil over the leg using a good brush and then rub Bay salt into the leg; this will make a lovely crisp crackling.

There are two ways of cooking this. You can either put a net around the leg to keep the shape and then place in a cooking bag, or you can roast just with the net — the choice is yours.

Cooking bag

To use the cooking bag method, place a net around the leg; this is to keep its shape. Then place in a cooking bag, making sure all the air is extruded, and tie the end with a piece of cotton string. Place in a large roasting dish — if roasting more than one, make sure they are at least 6" apart.

Before placing the leg in the tin, put a fine layer of salt in the bottom of the tin; this will stop the bag sticking to the bottom. The oven should be heated to 400°F. Put the meat into the oven and leave at this temperature for 30 minutes to seal the meat; then reduce the temperature to 375°F and cook at 20 minutes per pound.

When the cooking time has nearly finished, remove from the oven, remove the bag, and return pork to the baking tin. Sprinkle with some caraway seed; this will add more flavour. Return to the oven and flash off for 30 minutes.

Remove from the oven and transfer to a clean platter or a rack to drain off the juices and to let the pork cool down. When cold, wrap in greaseproof paper — do not vacuum pack this.

Roasting on salt

The other method of cooking is to lay the leg of pork on a fine layer of salt in the baking tin. Put into a hot oven (400°F) for 30 minutes to seal, lower the temperature to 375°F and cook for 20 minutes to the pound. 30 minutes before cooking has finished, remove from the oven, sprinkle with caraway seeds and return to the oven to flash off. This will ensure that the rind

is lovely and crispy and that you have a lovely piece of crackling.

Larding a leg of pork

A roast leg of pork can be a bit too dry when cooked. To counteract this, the meat can be larded with pork lardings (see page 67). As a personal preference, I would not use larding bacon as this will colour the pork.

Shoulders can also be cooked in this way. Many sandwich producers use the shoulder, and you can also use a middle of pork. The way to use a middle is to use a meat hammer to flatten the middle of pork. Make it very flat, spread the stuffing, roll up and tie at frequent intervals. This would be difficult to put in a cooking bag as it is very long.

Roast middle

Place it in a long tray with a layer of salt in the bottom. Cook at 20 minutes to the pound, and remove 20 minutes before cooking is complete. Sprinkle with caraway seed, return to the oven and flash off. Now remove from the oven and put on a rack to drain.

Do not leave it in its own juices as this makes it greasy. You will find a middle of pork slices very well, especially with the stuffing in the middle, and you will find this makes a very good display article.

Many manufacturers boil the pork first and then flash off in the oven, but I have never subscribed to that as it makes the pork soggy. I recommend the old-fashioned way.

Sage & Onion Stuffing

For your roast pork, use a sage and onion stuffing. Here are two recipes. The method is easy: mix the ingredients together and apply as described previously. When making these stuffings, do not use fresh onions but kibbled or dried.

Recipe 1

10lb fine-grade rusk

4lb kibbled (dried) onions

2lb sage, finely rubbed

1½lb rock salt, finely ground

Recipe 2

10lb medium-grade rusk

4lb kibbled (dried) onions

2lb sage, finely rubbed

2lb rock salt, finely ground

¾lb ground white pepper

10oz mace

6oz ground nutmeg

MAYNARD'S TIP

If you find the stuffing is a little dry when mixing it, add a small quantity of boiled sterilised water — this will give you a more pliable mixture. The recipes can be cut down to smaller quantities as desired.

Tongues

In my opinion, these are one of the easiest cooked meats to produce and they will give you good financial return. A display of fresh boiled tongue is an eye-catcher and the rewards are tremendous. They are so easy to make and will keep a long time. You will build your reputation and you can establish a very good trade around this product.

There are three types of tongue: ox, pig and sheep tongue. You can work pig tongue into small products, and you can do the same with lamb tongue, but in my experience the best tongues to produce are ox tongues. I found cow tongue a bit tough, so I always tried to use ox tongues.

I remember the best tongues I ever used, and you may not believe this, but they were reindeer tongues from Russia and they were fabulous. Tongues come from all parts of the world — Brazil, Australia, America, Canada — but I always tried to source my tongues locally as I thought it was the best thing to do.

You can produce four types of tongues:

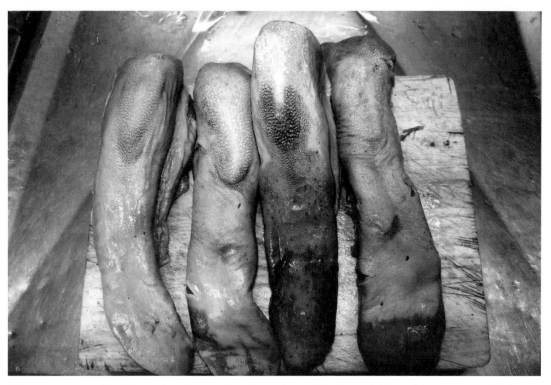

Raw tongues: the skin colouring varies from breed to breed. The skin comes off after the boiling stage of the process, so a dark pigmentation will not show.

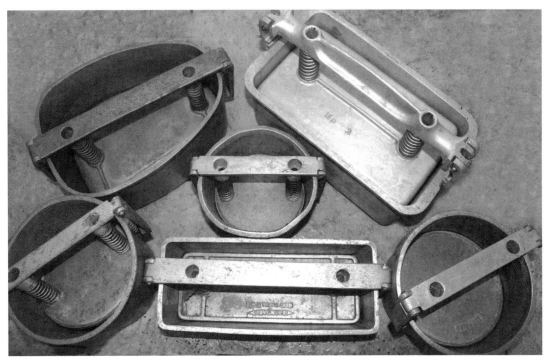

Presses of various shapes and sizes (clockwise from top left): pear-shaped press, brisket press, tongue press, pate press (bottom centre) and other circular tongue presses (bottom left and centre)

a smoked tongue, a sweet cured tongue, a spicy tongue and a garlic tongue. Which you produce depends on your trade and if you want a variety of tongues on your counter. You will find a niche in the market for tongue, especially garlic tongue.

Cleaning tongues

I will start with the cleansing pickle that will thoroughly clean the tongues and start the process of curing. It is made from the following ingredients, which you prepare as described on page 35:

10 gallons of water

10–15lb salt

4oz saltpetre

Before using the pickle, first take your tongues, place them in a container, and cover with large-grain salt. Leave for 3 or 4 hours to remove the slime. Then remove the tongues from the salt, rinse with water under a running tap, and make sure they are thoroughly clean. Now place them in a container with the cleansing brine and leave for 3 or 4 hours. This will start the osmosis which is the beginning of the curing process and it will also sterilise them a little bit longer, so you can be sure you are starting with a clean product.

Spice Brine
(for spicy tongue)

Once you have cleaned your ox tongues, they are ready for **curing**. This recipe is for a spice brine which you will find very pleasing to use.

Ingredients

> 10 gallons of water
>
> 20lb sea salt
>
> 8oz saltpetre
>
> 8oz Demerara sugar
>
> 3oz juniper berries
>
> 2oz pimento
>
> 3oz coriander

MAYNARD'S TIP

If you want a lighter cured tongue with gentle flavour, I suggest you pump with a 30% density and then put them in a 70% density brine.

Ox tongue, having been boiled, is placed in the press. Several more will be added before the gelatine is poured in and the lid fastened

Method

Boil the water to sterilise. When cool, add the salt and mix thoroughly. Boil the saltpetre in your dedicated saucepan. Leave to cool and then add to the brine. Put the sugar in a saucepan and heat until dissolved; leave to cool and add this to the brine. Grind finely the juniper berries, pimento and coriander, and put the mixture in a saucepan. Heat until infused; leave to go cold and add this to the brine, stirring well.

The product we are making depends very much on eye appeal so we have to make sure the colour is correct. Saltpetre brines are notorious for not producing a good colour so I suggest you add ¼oz of sodium nitrite. Heat this in a saucepan until dissolved, leave to cool and then add to the brine. As an alternative to sodium nitrite, add a small quantity of mature brine, but make sure you boil this first to sterilise before adding it to the main brine. Either of these methods will improve the colour of the product.

Once the brine has cooled, stir thoroughly; then take your brineometer and measure the density of the brine. These tongues need curing at a density of 70%. Once you have achieved this density, use a sterilised stainless steel bucket to remove a quantity of the brine and commence the cure.

At this point, all the small bones and debris should have been removed from the

Slices of tongue reflecting the curved edge of the press in which they were made

tongue. It is important that the tongue is smooth and pleasing to the eye. There is no point in curing ragged tongues.

I recommend that you begin the cure with the pumping method, but this is optional. Pump at 70% density. Pump from the back until they expand a little, but not too much because you will destroy the tissue; judging the time comes with experience.

Having done this, put them into the brine tank and leave in the tank for 4-5 days; if a light cure is needed remove on the fourth day. When cured, remove from the tank, wash with tepid water and dry.

The next stage is important. Some producers process the tongues at this stage, but I do not do this as we are after flavour and quality, so we will take the quality road. The way to do this is to put the tongues in a vacuum bag. Pack them one way then the other, and pack them tightly. The reason we are doing this is that we are after flavour.

Put them in the fridge and leave them to mature for about a week. By this time, all the herbs, sugar and nitrite will have blended to produce a lovely flavour and this method is the proper way to achieve this.

169

Cooking Tongues

A lot of people put tongues into a cooking net, but I always used to put them loose in the boiler. I would heat the boiler up to 185°F — it is important to check the temperature before placing the tongues in the boiler.

Put the tongues in and cook them for 4 to 5 hours. When cooked, fetch them out of the boiler but retain the hot water in the boiler.

Place the tongues on a stainless steel table and have two buckets of water ready on the table, one hot and one cold. The reason for the cold water is that the tongues are very hot to handle, so plunge your hands in this cold water to keep them cool. The hot water is for your knives. Put them in to heat the blades up: they will cut much better.

The next step is to take the skin off the tongues and to remove any other unsightly pieces of debris.

Now take your tongue press (which has previously been sterilised), warm with some hot water, and pour a little gelatine (see below) into the bottom. It is essential that you wash out the press in hot water before you pour in the gelatine, because if the container was cold the gelatine would set straight away which you do not want.

Place a tongue in the bottom of the press, and curl it so that it is completely curled around; if it is too big, cut the end of the tongue and save it. Then take the next tongue and place it the opposite way to the first one. Carry on doing this until you have more or less reached the top of the press where the notches are.

Then gently pour in the gelatine. Leave for half an hour; then go back and top up again. Place the lid on top of the tongues very gently. Now leave the press to cool. If you are using more than one press, leave a space between each one so that air can circulate and heat is not transferred.

When cooled, lower the lid of the press down a notch; this will make the tongues firmer for slicing. Place the press in the fridge and leave for 24 hours to set. Now remove the press and dip it quickly in hot water to loosen the tongues.

Turn the tongues out onto a wooden board. That is the easiest way to release them, but it is essential that you use a wooden board for this as a stainless steel table can chip the press.

Now wrap them in greaseproof paper, vacuum pack them, and leave in the fridge for 3-4 days to gather flavour.

MAYNARD'S TIP

The old Curers taught us to put a slice of lemon in the bottom of the press when cooking tongues. We also put bay leaves at the top and bottom. This would give a slightly different flavour.

Gelatine
(for pressing tongue)

For pressing tongue, you will need 2oz of powdered gelatine and 1 pint of water. Put 2-3 tablespoons of the water into a cup and sprinkle the gelatine into it; stir until the gelatine swells and absorbs the moisture.

Warm the rest of the water so that it is hot but do not boil, and stir the gelatine into it until it dissolves.

In my day if you wanted the gelatine to have a little colour, you added ¼oz of natural food colouring to achieve a slight pink. But today no dyes are allowed in cooked meats.

Smoked Tongue

The basic procedure for curing is the same as for spicy tongue, but this tongue is smoked and the following ingredients should be used for the cure:

    10 gallons of water

    20lb salt

    5lb Demerara sugar

    6oz saltpetre

    ¼oz sodium nitrite

When you have cured and matured the tongues, remove them from the vacuum bag and make sure they are dry.

Then put them in the smoke house and smoke at 110°F. Put them in the smoke house in a net or a muslin cloth depending on the type of smoke you require — light or heavy.

Once smoked, fetch them out and follow the procedure above for cooking, pressing, and releasing them. I suggest that if you are doing a large production, mark the presses of the smoked tongues and keep them separate from the other tongues.

Sweet Pickle Brine
(for sweet cured tongue)

Ingredients

    20 gallons of water

    15lb Bay salt

    5lb Demerara sugar

    1lb juniper berries

    2oz cloves

    3oz ginger

    8oz peppercorns

    ½oz bay leaves

Method

Follow the method for the spice brine, as explained above. Grind the juniper berries, ginger, cloves, peppercorns and bay leaves, and add this mixture to the brine. Before you use it, leave for a day to increase the flavour. Apply the brine as previously described.

Garlic Brine
(for garlic tongue)

Ingredients

> 10 gallons of water
>
> 20lb sea salt
>
> 8oz saltpetre
>
> 2lb Demerara sugar
>
> 4oz coriander
>
> 3oz ginger
>
> ½oz garlic
>
> ¼oz sodium nitrite

Method

Follow the method for the spice brine, as previously described. Crush the garlic in a garlic press and add this last to the brine, this will give a very distinctive garlic flavour. Use this brine as previously instructed. I recommend that you keep your garlic brine separate. Label well and cover with a lid to prevent the flavour transferring to other brines. Also, keep separate presses for your garlic tongue because the taste of garlic will linger no matter how well you wash them.

Pork Tongue

You now have recipes for four different types of ox tongue and now I will give you a recipe for pork tongue. I always work on the theory that if you have one product you will sell one product, but if you have three you will sell two. So now I am giving you five, and the choice is yours.

Start the same as with the ox tongues by placing the tongues in a cleansing brine. Clean them thoroughly, removing the slime and all the gullet bones. Wash them off with clear water and put in a tub of water and salt to start the osmosis stage of the curing process.

I suggest you do not pump these as they are only small tongues. Leave them in a 70% density brine for a week. Fetch them out, wash off and put them in a net. You should manage about 10 to a net. Cooking will be easier in a net as, being small, the tongues are difficult to remove from the boiler.

Heat the boiler to 185°F and cook for about 3-4 hours. Fetch them out, put them on your stainless steel table and remove all the skin. Use a smaller press for these tongues.

Wash it out with warm water and pour in a little gelatine, as explained above. (The gelatine would set straight away if the container was cold, so remember to warm the press first.)

Skin the tongues as previously described and pack into the press, putting the lid gently on top. When cooled, put the lid on a lower notch to compress the tongue. Put in the fridge and leave for

24 hours to set. Remove from the press as previously described, using a wooden block to turn them out. (Remember that using a stainless steel table can chip the moulds.)

Now proceed as before — vacuum pack them and put in the fridge. These tongues should be left for 3 days to gather flavour.

Tongue Sausage

The tongue for this sausage should be cured and cooked prior to manufacture. I always cooked the tongue prior to use as I believe this gave the sausage a longer shelf life — some manufacturers used to put it in raw but I preferred my way. In any case, it is no longer legal to add raw tongue to a sausage. You can no longer add offal of any kind to meat products unless the whole product is cooked prior to sale, under current UK legislation.

The tongue can be smoked or plain according to taste.

However, I suggest you do not make a vast amount of this sausage as the shelf life is limited. You will find this a very spicy sausage.

Ingredients

    15lb pork

    5lb back fat

    5lb cooked tongues

    4lb rusk

    5lb Bay salt, finely ground

    1oz ginger

    ½oz cloves

    ½oz cayenne pepper

    ½oz coriander

    1lb white pepper

Method

Prepare the tongues as previously described. Cure them using one of the recipes above; cook them as I have explained; then mince on a very fine grid. Put the pork through a medium grid, and put the back fat through a large grid.

Put the pork into the bowl chopper first, then the tongues; then put the seasonings on the top. Mix until the desired consistency has been achieved. Now add the rusk.

If the mixture feels tight, add a small quantity of water. Add the back fat to your chosen consistency — I suggest a very fine mix. Remove from the bowl and stuff out into sheep casings.

MAYNARD'S TIP

Always test your gelatine mix for strength by placing a little of it on a plate in the fridge. Leave for 10 minutes and if it has set well the strength is good. If not you know to add more gelatine.

Brawn

This is a product that has been neglected but I think it is a very profitable one to produce. I think the reason it has gone out of favour is that it has not been made correctly. When I was serving my apprenticeship, we used to make 2½ tons of brawn a week; this kept the factory clear of all the heads, trotters and tongues.

We had one concession: we were allowed two pots of brawn a day for our breakfast. We used to cut it thick on crusty bread, and it was the most delicious brawn I have ever eaten. When nine o'clock came, and after you had had an early start, it was like dining with the Gods and we really enjoyed it.

The brawn was a profitable product as we could utilise all the leftover pieces from whatever we were making. By producing brawn, you were turning waste into a useful product, and making more profit from this than from ham and sausages.

The old Curers told me that in the eighteenth century there were shops that sold only brawn and sausage but made an excellent living.

Brawn was predominately used in the north, being used for sandwiches, for salads as a slicing brawn, and for breakfast with two poached eggs and eaten with slices of toast; if you had brawn, you could always produce a meal.

It was a very useful product so I suggest you give this a thought because if you can turn waste into food you are on your way to success. I will tell you how we produced it, because in my opinion it was the best brawn I ever tasted.

I will also give you different recipes for the seasoning, as brawn varied from county to county and from person to person. I will also give you recipes for tongue brawn and tomato brawn and then you can choose which one you want to use.

There are many ways of presenting brawn. You can put brawn into a small container and sell it as pound pots; you can put it in a long container and slice it; or you can slice it and vacuum pack it for sale as a ready-to-use product.

Method

Use fresh heads. Cut off the eyelids, trim the ears, remove the eyes, and clean thoroughly. Using a blowtorch, singe all the hairs off and scrape with a knife to remove. Drop the heads into a salt solution to remove debris and blood, and leave until thoroughly clean.

Next take the pigs' trotters. Put your knife between the trotters and remove any debris. Singe any hairs off, and make sure all the hair is removed as there is nothing more off-putting than finding hairs in your food. Take your time to prepare this product and if you start right you will have an excellent return for your labours. Put the heads and trotters in the salt solution and you will find this will get rid of all the debris.

Brine for Brawn

The next thing to do is to make a brine. You will need a density of about 60%. Pick the brine you would like to use — you can use a sweet brine or any brine of your choosing as long as the density is about 60%.

To make your brine, follow previous instructions. Just to remind you, the proportions are roughly 10 gallons of water, 20lb of good quality salt (such as rock salt or Bay), and 7oz of saltpetre. As I said before, boil the water, add the salt, boil the saltpetre separately and add that. To produce the required colour, it is always a good idea to add an older brine, having boiled this before adding it to your new brine; this will produce the colour.

Once the brine has been made, put all the trotters and head pieces into this brine for 7 days. Remove and put in the boiler. Fill the boiler with cold water, then flush the water away. Repeat this three times — this will flush away the scum and all the odd bits of bone and blood. Now fill the boiler with fresh water. The water needs to be about 12" above the heads and trotters so that they are completely covered when cooking, but do not fill the boiler completely; three-quarters full should be sufficient.

Cooking the Brawn

Light the boiler using a low heat. Cook this product slowly — if you cook it too fast the calcium leaches from the bones and the brawn will be cloudy.

When the heads and trotters are well cooked, the meat should be falling off the bones. Remove the bones and meat and place on a tray, leaving the liquid in the boiler; the liquid is important as this will make up your stock. Remove the rind and keep this separate. Then remove all the meat from the bones and discard the bones. When you have cleared all the gristle and bones, mince the meat using a large grid and put this into a container. That completes the first stage of cooking.

Cooking Rinds

For the next stage, you should have a second boiler. Use this one for cooking the rinds. There are two ways of doing this: you can have your rinds loose or you can put the rinds in a cooking net. The cooking net is a good idea as it keeps the rinds together and they are easier to remove from the boiler.

Cook the rinds until they are tender. Take them out of the boiler, mince them on the finest grid you have, and place them in a container. Retain the fluid in this boiler — this is important. You have cooked the rinds in this boiler and there will be a lot of gelatine left behind in the liquid, which will help to make the brawn.

The next thing to do is to put the meat in the boiler: use either boiler, and add liquid from the other to make up stock as required. Add the minced rinds and mix well, as there will be a lot of jelly in this mixture. Put this mixture on a slow heat. You must stay with this stirring frequently, as the brawn will stick to the bottom if left. A lot of brawn will be lost if the mixture is left to simmer, as the gelatine will stick to the bottom of the boiler and burn. Keep stirring and the brawn will start to boil up. Now put your seasoning in (see the recipes below) — it is a good idea to taste the mixture to check the flavour. At this stage, many people add lemon juice to the mixture; this is optional and you may not want to do this.

Adjustments

By this time, the brawn should have reached a good consistency. If you feel it hasn't then add some gelatine to the mixture, about 2-4oz, stirring this well. Once you have done this, check the colour; it should be a nice pink colour. In the old days if it lacked colour, we simply added a small amount of food colouring: this is no longer legal.

Raise the temperature up to 175-180°F, stirring all the time. It is important to keep stirring to reach the desired consistency and to prevent the gelatine from sticking on the bottom.

Pressing

Have all your containers ready — this should have been done previously. Before you put the containers on the table, it is a good idea to put greaseproof paper on the table and place your containers on the paper. Then if you spill the brawn it will be easier to clean up.

Using a jug, scoop up the mixture and start filling the pots and containers; these should have been thoroughly cleaned prior to use. Do not fill the containers to the top.

After an hour, stir each container to distribute the gelatine. This will give you a unified product as gelatine has a tendency to drop to the bottom, so do this a couple of times. Once cooled, place them in the fridge, marking each container for which purpose it is intended, such as slicing brawn or small pots for individual sale.

Tomato and Tongue Brawns

I will now give you two variations which will increase your menu. The first is **tomato brawn**. To make tomato brawn, follow the same process but sieve some tomato puree and add this to the mixture.

The second is **tongue brawn**. This is made by curing the tongues in a sweet pickle first; then you remove the skin, cut them into ½" squares and add them to the brawn mix.

Brawn Seasonings

I've collected these seasonings from different areas of the country and you must decide which one you prefer — the choice is yours.

Seasoning 1

    7lb fine salt

    5lb white pepper

    2oz nutmeg

    2oz mace

    1½oz ginger

Seasoning 2

    5lb fine salt

    3lb white pepper

    1oz cayenne pepper

    1oz cloves

    1oz ginger

Seasoning 3

    7lb fine salt

    3lb white pepper

    1oz nutmeg

    2oz mace

    2oz cayenne pepper

    ½oz ginger

Seasoning 4

    10lb fine salt

    6lb white pepper

    2oz mace

    2oz ginger

    ½oz nutmeg

Seasoning 5

    6lb fine salt

    2lb white pepper

    ½oz cayenne pepper

Seasoning 6

    10lb fine salt

    4lb white pepper

    1oz nutmeg

    2oz mace

    1½oz cayenne pepper

    1oz cinnamon

Savoury Ducks & Faggots

These are cousins. The difference lies in the seasonings and the mincings: some are produced on a fine grid and some on a large grid. I think you should consider making these products as they are becoming very popular and there seems to be a ready sale for them.

Making savoury ducks and faggots is also a way of using all the by-products of the practice as well as keeping the factory clean of bits and pieces.

Savoury ducks consist of hocks, head meat, liver, and any oddments; there are no limitations. I will give you a general recipe for savoury ducks and you can add or subtract items as you wish.

Savoury Ducks

Ingredients

20lb rinds & liver (well cooked)

8lb large-grade rusk

1lb rice flour

1lb suet

2 onions

caul fat

seasoning (see recipes)

Method

Cook the rinds, liver and oddments until very tender; then mince on a fine grid. Put the minced meat in your bowl chopper. Mince the onions and add to the bowl. Mix the rusk with stock from the boiler and add that to the bowl chopper. Add the seasoning (see the recipes); add the suet and the rice flour. Mix well until you have a pliable mixture.

The next stage is to flour a pastry board and roll the mixture into a cigar shape. Now take 4-6oz of the mixture and shape it into a ball. Wrap the caul fat around each one, and place in a baking tin in rows.

On the bottom of the tin pour some stock from the boiler; this stops the ducks from sticking to the tin and burning. Place in the oven and bake at 400°F for 1-1½ hours.

When cooked, remove from the oven and leave to cool. When cold, put the ducks on a clean platter — do not move them whilst hot as they will break. These are your savoury ducks — serve them with mushy peas, gravy and crusty bread, and you will have a dish fit for the Gods.

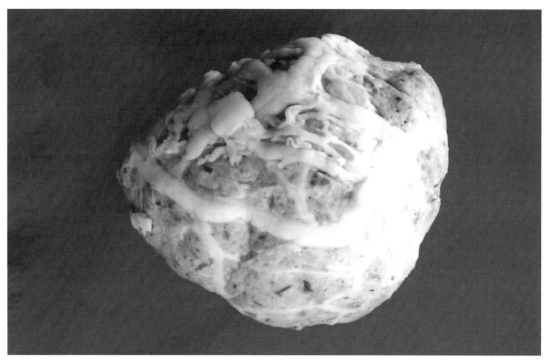

A savoury duck wrapped in caul fat and ready for baking

Seasoning for Savoury Duck Recipe 1

Use ½oz to 1lb of meat. If you have some salty products in your meat mixture, decrease the amount of salt in the seasoning.

    10lb fine salt

    3lb black pepper

    3oz sage

    2oz marjoram

    1oz savoury

    2oz cayenne pepper

Seasoning for Savoury Duck Recipe 2

Use ½oz to 1lb of meat. If you have some salty products in your meat mixture, decrease the amount of salt in the seasoning.

    5lb fine salt

    1lb white pepper

    6oz sage

    2oz nutmeg

    1oz marjoram

Faggots

Ingredients

24lbs assorted meat products, including fresh hocks, pork hearts, liver and other oddments

6lb medium-grade rusk

1½lb Spanish onions

caul fat

seasoning (see recipes)

Method

Cook the hocks, pork hearts, liver and oddments. Then put them through a medium grid and place in your bowl

Faggot seasoning – Recipe 1

3lb fine salt

2lb black pepper (ground)

3oz cayenne pepper

2oz sage

1oz thyme

Use ¾oz to 1lb of meat. Review your seasoning if salty meat has been used.

Faggots, ready for the oven

180

chopper. Mince the onions and add to the bowl. Mix well, and add the seasoning and rusk. Add a small amount of sterilised water, and mix thoroughly until the desired consistency is achieved. Remove from the bowl chopper and place on a floured board. Shape into balls, using approximately 4–6oz for each ball, and then wrap in caul fat.

Place the faggots in a baking tin, putting a small amount of water in the bottom to prevent sticking. Bake at 375–420°F for approximately 1½–2 hours. The cooking time is longer than for savoury ducks as the oddments have not been cooked.

Follow this recipe and you will have lovely flavoured faggots. Eaten with mushy peas, gravy and crusty bread, they will give you a filling and enjoyable meal.

Cutting off the trotter from the hock: the latter an essential ingredient for faggot-making

Faggot seasoning – Recipe 2

5lb fine salt

3oz sage

2oz savoury

2lb white pepper

Use ¾oz to 1lb of meat. Review your seasoning if salty meat has been used.

Pressed Pork

This is a product that is not used very much today, but in the past it was very popular. It is a very easy thing to make, and a good outlet for shoulders and hocks. It looks attractive and is very succulent.

Spice Brine for Pressed Pork

Ingredients
    20 gallons of water
    14lb rock salt
    6lb light sugar (light Muscovado or Demerara)
    1lb juniper berries
    1oz cloves
    8oz saltpetre
    1oz pimento
    8oz peppercorns
    ½oz bay leaves (ground)

Method
Boil the water; then add the salt. Boil the saltpetre in a separate saucepan and leave to cool; then add to the brine. Boil the sugar until dissolved; leave to cool and then add to the mixture. Grind all the herbs and spices including the peppercorns. Boil in a saucepan, leave to cool and add to the mixture. Leave the spice brine for a day and then test for density; you do not want this brine to be too salty so the density should be 60%.

Pressed pork is made in a square mould, and I will go through the different stages.

It is essential you have a special brine for pressed pork, ideally a small quantity of very mature brine mixed with the new brine which will give the pressed pork colour and a bit of character (see the recipes below).

Strong colour is important, as grey or dark pressed pork does not look very appetising. It is therefore advisable to use a mature brine and, if you want to hasten the process, the knack is to use a small amount of brine that you already have and add this to the new brine; this will hasten its maturity.

Plain Brine for Pressed Pork

Ingredients
    10 gallons of water
    20lb salt
    6oz saltpetre
    5lb light sugar (light Muscovado or Demerara)

Method
Follow the procedure for the spice brine, as described above. This brine also needs maturity, flavour and colour, so use the above tips for achieving this. You will sell this product on eye appeal and colour is very important.

Making Pressed Pork

To make pressed pork, first take the shoulders and remove the rind and any excess fat. Pump the shoulders and also the hocks with a 60% density brine (see the recipes); then put in your chosen brine for about 6 days. Remove, wash off, and then put in the boiler. Cook at about 180°F until tender so that the bones can be easily removed.

When cooked, fetch from the boiler and place on a table. Remove all the bones and gristle. Have ready your square press, which has been sterilised with hot water. Warm the press with hot water as this prevents the meat sticking to the press. When packing the meat, arrange in layers until you nearly reach the top.

The next stage is to boil a pint of water and mix some gelatine into it; then pour the mixture into the press. It is a good idea if colour is needed to add a small amount of food colouring to the gelatine. Leave for an hour, then add some more gelatine. Put the lid on tightly so that the pork is compressed.

When cold, put in the fridge for a couple of days. Fetch out; then dip the press in hot water so that the meat will come out easily. Wrap it in greaseproof paper and vacuum pack it. Store in the fridge until required.

Pressed Ham & Tongue

This is another old recipe. Sometimes in production there is a surplus of ham hocks and bacon hocks, so you may find this recipe useful. If you have an excess of pork tongues, you can use these as well, and you can sell the product as a ham and tongue slice.

I suggest that you wash off the hocks first; then put them in the boiler and cook at 180°F until tender. Cook for 3-4 hours until the bone comes out. Boil the tongues with the hocks; if a sweeter product is required, add a small amount of honey to the water.

For pressing, use a square or round mould. Make up some gelatine as previously instructed, and put a layer in the bottom of the press; this gives stability. Lay the hocks and tongues flat in the press and build them up, placing them alternately in the press. Fill the press until the top is nearly reached. Now pour the gelatine over the pieces; leave for an hour then top up. Put the lid on and leave to cool.

When cold, put the press in the fridge for 2 days. Remove from the fridge, and wrap and vacuum pack as described before.

Luncheon Meat

This is something that, in my opinion, is due for a revival. Luncheon meat has become unpopular but I think this is because it has been made with poor quality materials. At one time it was a very popular and profitable product, and in my experience there are many advantages to producing a good quality luncheon meat.

It gives variety to your menu, it has a long shelf life and it is very versatile. It can be eaten cold with salad or on sandwiches, or you can fry it and have it with egg and bacon.

In addition, you can make luncheon meat from a range of different meats and with different seasonings (see the recipes). I think it is a tremendous product that has been neglected and I hope this book will encourage people to produce it. I think you will have a good return on your labour and materials if you do.

Luncheon meat can be cooked by putting it into large beef bungs (casings) – synthetic or natural – or some people choose to pack it into square containers. Personally, I never liked square containers as they always seemed to leak. The juices of the luncheon meat would escape from the container into the boiler, so I was never a great believer in containers; I always used natural or synthetic casings, which trap the natural juices much better.

Then, after cooking, you can remove the skins and dress the luncheon meat with a ham dressing. This looks very attractive on a counter, and I do hope you consider this way of making it.

When curing luncheon meat, it is essential to have a stable colour; sometimes with brining meat we don't achieve this. The method I am going to suggest is, I think, foolproof. Using this method for all the luncheon meats (including the smoked, the beef and the pork varieties) will give you a better product and I think you will agree.

Curing Luncheon Meat

First the fat: A week before making your luncheon meat, I recommend that you first cure the fat you will be using. To do this, make a mixture of 20lb salt and 6oz saltpetre. Take some large pieces of back fat, cover with the salt and saltpetre mix, and leave this for a week. It is best to cure the

fat separately and in advance of the lean meat because you really want the fat to be quite hard and to stand firm and distinct in the final mix. By giving it a longer cure you will be taking out more of the moisture and you will achieve this firm quality in the fat. The same method applies for salami-making.

Second the lean meat: You have already cured the fat so the next cure (see blue box below) is for the lean meat only. Spread the mixture evenly over the meat, cover with a piece of polythene and put in the fridge. Leave for 24 hours. Fetch the lean meat out and see if the cure has taken. If you are not satisfied, return it to the fridge and leave a little longer. When you are satisfied, remove it from the fridge.

Dry cure for the lean meat

Ingredients
6lb fine salt
8oz light sugar
½oz sodium nitrite
4oz finely ground saltpetre

It is essential that you mix these ingredients really well because you want the cure to be evenly applied. You will need to use it as ½oz cure to 1lb of lean meat.

Preparing and Cooking Luncheon Meat

When the meat has been cured, the next step is to weigh all your ingredients and have your binders ready. Put the meat in the bowl chopper, add the seasonings and turn the bowl chopper for a couple of revolutions or so until the meat and seasonings are well mixed.

Now add the binders and mix until absorbed. If the mixture is too stiff, add a small amount of sterilised water. Mince or dice the cured fat; then add to the mixture. Do not chop the fat too fine, but chop into small chunks as this will keep the product moist. Mix until the desired consistency is reached.

The mixture is now ready for the sausage filler. Using natural or synthetic skins, fill the skins firmly, but not too tightly or they will burst in cooking. Tie one end of the skins, and leave a loop for hanging up. When the casing is filled, twist the end twice and loop around; then place on a tray. Stuff out the rest of the mixture, and place all the luncheon meat sausages in the tray ready for cooking.

The boiler should have been previously filled and the temperature heated to 180°F (it is worthwhile investing in a thermometer). Put the luncheon meat in the boiler. When you do this the temperature will drop, so heat the boiler back to 180°F and keep it at this level — this is important, as the centre of the luncheon meat must be cooked. The meat will take 2½–3 hours to cook at this temperature, depending on the diameter of the casings.

Once cooked, remove from the boiler

and spray with cold water to bring the temperature down; then place in a bath of cold water to cool completely. When cold, remove and hang up to dry, using the loops already made. Leave a space of 6-8" between them so the air can circulate. When cold, put in the fridge and use as trade demands.

Luncheon Meat Recipes

There are a number of different binders you can use, including plain flour (this tightens the mix), cornflour (this moistens the mix), and medium-grade rusk (this will absorb the meat juices during cooking).

These three are excellent binders for luncheon meat and I suggest you use them all, combining them in the ratio of 2:2:1 (40% plain flour, 40% corn flour and 20% rusk).

This combination will give the luncheon meat a good cutting edge while keeping it moist. When you add water to the combination, do this slowly.

Another good binder is farina. I suggest you use 1lb of farina to 1 pint of water. Mix the farina to a paste-like consistency and add to the mixture after the seasonings.

Do not overload the product with binders as the natural flavour will be destroyed, so do keep them to a minimum. The binders are there to retain the natural juices, not to bulk out the product.

Luncheon meat can be smoked for 3-4 hours and this will give you another variation with a pleasant smoky flavour.

Pork Luncheon Meat

Ingredients
> 15lb lean pork
>
> 4lb cured back fat
>
> 3lb farina
>
> 2lb medium-grade rusk
>
> 6 pints of water (boiled and sterilised) – adjust as needed
>
> seasoning (see blue box)

Method
Prepare and cook as described above.

Seasoning for Pork Luncheon Meat

Ingredients
> 8lb white pepper
>
> 8oz mace
>
> 4oz nutmeg
>
> 4oz ginger
>
> 4oz coriander

Use ½oz of seasoning to 1lb of meat. There is no salt in this seasoning as the salt has already been added as part of the back fat curing process.

186

Bacon Luncheon Meat

Ingredients

20lb bacon trimmings

10lb bacon fat (remove the rind prior to use)

3lb farina

3lb medium-grade rusk

7 pints of water (boiled and sterilised)

seasoning (see blue box)

Method

Prepare and cook as described above. I recommend this recipe as a very good way of keeping the fridge and the factory clear of any leftover meat!

Seasoning for Bacon Luncheon Meat

Ingredients

10oz white pepper (freshly ground)

5oz nutmeg

1½oz ginger

1½oz coriander

1½oz cayenne pepper

There is no salt in these ingredients as this has already been added to the bacon in the earlier curing process. Use ½oz of seasoning to 1lb of meat.

Pork & Beef Luncheon Meat

Ingredients

9lb lean shoulder pork

9lb lean beef

10lb cured bacon

8lb cured back fat

8lb binders

8 pints of water (boiled and sterilised)

seasoning (see blue box overleaf)

Method

For the binders, use a mixture of plain flour, corn flour and fine-grade rusk (40% plain flour, 40% corn flour, 20% rusk). Put the binding mixture in a container, add half the water and make a paste. Cure the beef, pork and back fat as described above. When cured, put the beef and the pork through a large grid. Put the bacon bits and the back fat through a very fine grid. Put the beef and pork in the bowl chopper. Add the seasoning and mix thoroughly; then add the bacon bits. Now add the binding paste and mix until evenly distributed. Add the cured back fat last with the remainder of the water. Put into natural or synthetic casings and cook as above. The ingredients do not need salt as all of the meat has been cured.

MAYNARD SAYS

Only foolish people throw money away, the wise pick it up.

Seasoning for Pork & Beef Luncheon Meat

Ingredients

4lb white pepper

2oz mace

1oz ginger

2oz nutmeg

2oz coriander

Method

Keep the seasoning in an airtight container, marked 'pork & beef luncheon meat.' Use ½oz of seasoning to 1lb of meat.

Smoked shoulder pieces for smoked luncheon meat

Smoked Luncheon Meat

Ingredients

20lb lean shoulder pork

10lb belly pork, lightly cured and smoked

4lb farina

4 pints of water (boiled and sterilised)

seasoning (see blue box)

Method

Cure the belly pork in a sweet cure and smoke lightly. When smoked, leave to go cold. Now remove the rind, dice into squares and put in the mixture; this will give the luncheon meat its smoked flavour. If a heavier smokey taste is desired, increase the amount of smoked belly pork in the recipe. Use the method for mixing as described above. I think you will find this a simple recipe to do, as you only have to smoke the belly pork and not the rest of the meat.

Seasoning for Smoked Luncheon Meat

Ingredients

14lb white pepper

1lb ginger

8oz coriander

8oz nutmeg

8oz mace

Put the seasoning in a separate jar and label 'Smoked luncheon meat seasoning' as the ingredients are quite different to the other seasonings you will use. Use ½oz to 1lb of meat.

Commercial Luncheon Meat

Ingredients

14lb beef

14lb lean shoulder pork

20lb cured back fat

16lb bacon

2lb rusk

2lb flour

1lb cornflour

12lb water (boiled and sterilised)

seasoning (roughly ½oz of seasoning to 1lb of meat, but this can be adjusted to taste)

Method

Having cured the fat and the lean meat, you put the beef and the pork through a large mincing grid. You then put the back fat through a finer grid and the same for the bacon bits.

Place the rusk, cornflour and flour in a bowl with half the water. Mix thoroughly.

Now put the beef and lean pork in the chopping bowl with the seasoning and mix thoroughly, adding the bacon bits. Now add the rusk and flour paste and mix further.

Lastly, add the back fat and some or all of the remainder of the water (adjust to make a good, workable sausage consistency) and mix. Now the mixture is ready to be packed into natural (or if you must synthetic) casings. For the cooking see instructions on page 21.

Seasoning for Commercial Luncheon Meat

Ingredients

8lb white pepper

2oz ginger

4oz mace

4oz nutmeg

3oz coriander

Mix these ingredients well and keep them in an airtight container for future use.

MAYNARD'S TIP

If you use natural casings and you want them to have an attractive colour, use a red or brown dye from the butchers' supplier. When removing the luncheon sausage from the boiler, dip it into the dye for a few seconds, then plunge into cold water with sufficient salt — this will set the colour on the casings.

Liver Sausage

There are many varieties of liver sausage and I will give you a number of recipes; you can then decide which one suits your needs. We used to manufacture a considerable amount of this product and the recipes will show you how we used to do it. The best livers in my opinion are pork livers; beef livers are too harsh and calf livers would be too expensive to use. Liver sausage is a good outlet for your heads, trotters, livers and tails. It is another unfashionable product, but I think this is because it is often poorly made and perhaps not marketed very well.

Traditional Liver Sausage

Ingredients

14lb belly pork

12lb pig's liver

7lb cooked oddments (e.g. head meat)

2lb farina

8oz onions

add a little cornflour or flour if the mixture seems a bit sloppy

seasoning (see blue box)

Method

For the meat oddments, you can use pigs' heads and trotters. First remove the lips, eyes, eyelids, and hairs from the heads. Make sure the hairs are removed — if necessary, use your blowtorch. Once the heads are prepared, place them in a cleansing bath consisting of a strong solution of salt water. Clean the trotters thoroughly, making sure they are well cleaned between the toes. Rinse off and put those in a cleansing bath.

The reason for adding the trotters is that they produce an abundance of gelatine and this will knit the product together. The pigs' ears and tails can also be used: clean them thoroughly and put them in the brine.

After cleaning, place all the oddments in the boiler. At this stage, some of the old Curers would add a teaspoonful of vinegar which was said to enhance the flavour, but this is optional. Cook all these oddments thoroughly. Remove from the boiler, leave to cool and then remove all the bones — this is very important as there must not be

---

Seasoning for Traditional Liver Sausage

Ingredients

10lb salt (finely ground)

2lb white pepper

3oz cayenne pepper

8oz ginger

2oz pimento

For the seasoning, use ½oz to 1lb of meat.

one bone left behind. Keep the water for stock. After removing the bones, put all of the oddments through a very large grid.

Prepare the liver by removing all the sinews and vessels. Place in salt water for an hour, then blanch in very hot water — this loosens the membrane around the liver for easier removal. Now cut into long slivers, coat with flour and fry to seal. Finely chop the onions, and cook in butter until transparent. Put the chopped onion and liver through a fine grid.

Now put the meat oddments, liver and onion in the bowl chopper, and mix gently until thoroughly blended. Add the seasoning, rusk, corn flour, and plain flour. If the mixture is too tight at this stage, remove some of the stock from the boiler, sieve and add this to the mixture.

While the mixture is turning, add the gelatine to the bowl chopper. When this is absorbed, add the belly pork; this should have been cubed to the desired size. Mix until thoroughly blended, remove from the bowl chopper and leave for 1 hour to settle.

Place the mixture into synthetic casings. Now place the liver sausage into the boiler and bring the temperature up to 180°F — the boiler should be three quarters full of cold water. Cook for 1½-2 hours; cooking time will depend on the diameter of the sausage. Once cooked, remove from the boiler, wash off with cold water and then leave in cold water to cool. When cold, remove and dry; then label with the date and time of production.

Fine Liver Sausage

The **ingredients** and the **seasoning** for this liver sausage are the same as the previous recipe. For the filling, use the **method** described above, but when mincing use the fine grid. Put the oddments in the bowl chopper and follow the above procedure until the desired consistency is reached. Pack and store as above. This liver sausage was used in mixed grills — the synthetic casing was taken off, and the sausage meat was sliced and served with egg, bacon, and oatcakes in restaurants, providing a good northern breakfast.

Smoked Liver Sausage

Ingredients

  15lb sheep's or pig's liver

  6lb cured belly pork (smoked or plain)

  3lb farina

  3lb rusk (medium grade)

  seasoning (see blue box)

Seasoning for Smoked Liver Sausage

Ingredients

  8lb Bay salt (finely ground)

  2oz pimento

  5oz freshly ground ginger

  4oz cayenne pepper

  2lb white pepper

  Use ½oz seasoning to 1lb of meat.

Method

When you make this sausage, remember that the pork has already been cured so the salt content of the recipe may need to be adjusted. Clean the liver thoroughly using a salt solution. Remove all the sinews and then blanch in boiling water. Mince the belly pork through a medium grid. Mince the liver through a fine grid. Now put the minced meat and fat in your bowl chopper. Mix the farina with a small quantity of water, then add to the bowl chopper. Mix well, then add the seasoning and mix thoroughly. Now add the rusk to make the product just stiff enough to slice. Stuff out in beef runners, and cook at 180°F for 40-50 minutes. When cooked, remove from the boiler and wash off in cold water until cooled. When cool, wipe off, hang on sausage sticks and put in the fridge to mature.

MAYNARD'S TIP

The cooking time of your liver sausage time depends on the diameter of the casing. The bigger the diameter, the longer it takes to cook; judging the cooking time comes with experience.

Bacon & Liver Sausage

Ingredients

17lb pigs' livers
25lb lean pork
10lb smoked belly pork
seasoning (see blue box)

Method

Prepare the meat as described above. Mince the smoked belly on a medium grid; do the same with the pork. Put the belly and the pork in the bowl chopper and mix thoroughly; then add the liver and the seasoning. At this stage, you can also add some gelatine to the mix — this has good binding qualities. If the mixture is too tight, use stock from the boiler as described above. Now follow the procedure as described in the previous recipe.

Seasoning for Bacon & Liver Sausage

Ingredients

3lb fine salt
1lb white pepper
2oz allspice
1oz coriander
2oz cayenne pepper

Use ½oz seasoning to 1lb of meat.

Polony Sausage

This is a forgotten delicacy. I really think this sausage has a great potential in the home and in the catering industry.

It uses all the leftovers in the factory, such as pork hocks, bacon hocks, rind emulsion, head meat, trotters — any of these items can be used in reasonable quantities.

The characteristics you are aiming for are, firstly, that it spreads well — polony is great for sandwiches.

Secondly, the texture should be very fine, so attention must be paid to achieve this while retaining the delicate flavour.

Ingredients

    20lb lean pork or oddments

    10lb fat

    3lb boiled rice

    3lb rusk (medium grade)

    seasoning (see blue box overleaf)

Polony sausage. Nowadays the casings are plastic. But the old method involved large hog casings, which would be dipped into a red polony dye (hot) with added salt, then the sausage would be dipped into cold water to fix it

Seasoning for Polony Sausage

Ingredients

10lb salt (finely ground)

4lb white pepper (freshly ground)

8oz coriander

6oz ginger

4oz cayenne pepper

4oz cloves

3oz paprika

Use ½oz seasoning to 1lb of meat.

Method

If using oddments such as pigs' tails or heads, these must be cooked prior to use and all bones should be removed. If using straight pork, this can be added to the mix without prior cooking. Mix the meat (lean pork or oddments, or a mixture of both) on a very fine grid. Put it in the bowl chopper, then add the seasoning and mix well. Add the rice and mix well; then add the fat and mix thoroughly. The last item to add is the rusk. This will tighten the mix — the mix must be very fine and very tight.

Now remove the mixture from the bowl chopper and leave to rest for 30 minutes. Stuff out into beef runners. That is the traditional way, but you can use synthetic casings — the choice is yours. Personally, I think they look very nice in beef runners, but if the trade demands the longer casings you may choose to use those.

When stuffed, place the sausage in the boiler and heat to 180°F. If using beef runners, cook for about 40 minutes. The synthetic casings will take longer, and you will need to use your own judgement.

When cooked, remove from the boiler and — if you are using natural casings — dip the sausage in polony dye (this is a special food dye that produces the red of the polony sausage). The dye should be kept at 180°F.

Once dipped in the dye, dip the sausage in a bath of salt and this will fix the dye. (Note that if you are using synthetic casings, there is no need to dye these, so leave them as they are.) Put the sausage in cold water to cool. They should be chilled quickly, and sliced on a horizontal slicing machine.

Saveloy Sausage

Saveloy sausages used to be smoked and then cooked, but to save a lot of unnecessary work I think it is better to incorporate some smoked belly into the filling.

Ingredients

6lb smoked belly pork

9lb pork

9lb beef

4lb fat

5oz rusk (medium grade)

seasoning (see blue box)

Method

Put the smoked pork, the lean pork and the beef through a medium grid. Put the fat through a large grid. Mix the rusk with water until it has a paste-like consistency.

Put the smoked pork, beef and lean pork in the bowl chopper and turn until well mixed. Add the seasoning and mix thoroughly; then add the rusk, followed by the fat. If you feel the mixture is too stiff at this stage, add some water. Remove from the bowl chopper and put in a container.

Leave for a while to settle; then stuff into large hog casings. Do not link, but twist into 8" lengths. To secure, tie at the ends. Now put in the boiler and cook for 30 minutes at 175°F — do not cook at a higher temperature as they will burst.

When they have cooked for half an hour, remove from the boiler and put on sausage sticks to cool. When thoroughly cool, put into a Brunswick dye or a polony dye to colour them, just as you do with polony — the choice of dye is up to you. Either way, if you want a coloured sausage, heat the dye to 170°F then dip the sausage into it. Plunge into cold salted water to fix the dye.

Seasoning for Saveloy Sausage

Ingredients

8lb fine salt

4oz sage

4oz ginger

3oz cayenne pepper

3lb white pepper

Use ½oz seasoning to 1lb of meat.

MAYNARD'S TIP

I never used to dye Saveloys as I thought they looked attractive as they were — the colour from the smoking gave them a lovely golden colour.

Pig's Trotters

One of the forgotten treasures of our trade. Most manufacturers discard pigs' feet as unwanted items, but there are those amongst us who feel we should utilise every oddment. So I say, think again and make a better job of it, because it is easy to throw money away.

Pigs' feet contain a large amount of gelatine which has a good binding quality and can be utilised in brawn, pâté, pork pies and most varieties of luncheon meat. For the straight sale of pigs' feet, I used the process that I am going to tell you about now, and I never had any problems in selling them.

I used to supply high-class restaurants as they became a delicacy and I would like you to consider this recipe as it will put money in your moneybox!

Method

First remove the nails from the feet. Next, use a blowtorch to remove all the hairs, and then thoroughly wash the trotters to remove any residual muck. They must be scrupulously clean. Once you have achieved this, drop the feet into a brine consisting of:

> 25lb Bay salt
>
> 10 gallons of water
>
> 6oz saltpetre
>
> 4oz light sugar (I suggest light Muscovado or Demerara)

To prepare the brine, boil the water, add the salt, grind the saltpetre, add the sugar and stir well. The density needs to be about 60%. It does not need to be too salty as this will destroy the flavour of the product.

The brine needs to be mature as we want the pigs' feet to have a nice colour. A new brine will not produce the desired colour. To overcome this, take a saucepan full of an old brine, boil it up and add it to the new brine; this will give it a start and produce a good colour.

When the brine is ready, place the pigs' feet in it, making sure they are thoroughly cleaned beforehand. Leave them in this brine for at least 8 days; this is to make sure that the colour goes right through the product.

After 8 days, remove from the brine and wash off. Traditionally, pigs' feet were placed in a boiler, but I found that the feet broke up when boiled at high temperatures, so I decided on a different method: I used to put a grid in the boiler and place the pigs' feet on this grid with the water below. This method steams the feet rather than boils them, and it takes about an hour for them to cook at 200°F.

When you are satisfied they are cooked, turn off the heat. Do not take them out yet as they will break, but leave them in the boiler to cool down first. Traditionally, some Curers would add vinegar to the water to enhance the taste but again this is up to you.

To reiterate, heat the boiler to 200°F. Steam the pigs' feet until you are satisfied they are cooked. Turn off the heat. Leave feet in the boiler to cool down. Take out when cold.

Now place the feet on trays. At this stage, you can add some malt vinegar and I found adding it at this stage improved the flavour. I found this was the best way to process pigs' feet and if you can build up a trade for them you will have a winner. I found the best way to enjoy trotters was with a crusty cob, butter and a glass of cider — there is no nicer meal.

Pork Pâté

This is a simple thing to make, and if you develop a trade for this product you will find it is very profitable. Pâté is a very popular dish for dinner parties, for sandwich fillings, and as an accompaniment to salads.

Ingredients

12lb belly pork, with the rinds removed

12lb pig's liver

1lb onions

12 streaky rashers, with the rinds removed

seasoning (see blue box)

Seasoning for Pork Pâté

Ingredients

2lb rock salt (finely ground)

8oz white pepper

6oz mace

6oz nutmeg

7oz ginger

Use the seasoning to taste; it is very difficult to specify the proportion per meat, as there is salt in the rashers and this can affect how much seasoning is required.

Method

Prepare the liver by soaking in salt water to remove blood and debris. Take out and dry; then remove the membrane and cut into slivers. Coat slivers in flour and fry in butter until sealed. Slice onions finely and fry. Mince the lean pork through a medium grid, and do the same with the belly pork. Mince the liver and fried onions through a fine grid.

Now put the lean pork, belly pork, liver and onions in the bowl chopper — add your seasoning last. Turn the bowl chopper until a fine paste is achieved. Stop turning occasionally and taste to see if the seasoning is sufficient; if not, add some more. When the desired consistency is reached, stop the bowl chopper.

The next stage is the preparation of the containers for your pâté. The choice of container will depend on your outlets: you may want to use ceramic, or a simple tin foil. First, prepare the streaky bacon by laying between two sheets of polythene. Using your meat hammer, flatten the rashers so that they are level. Place the rashers in the bottom of the containers so that the rashers come up the sides and hang over the top. Next, spoon the pâté into the containers and wrap the bacon over the top. Place the containers in a large baking tin with water a quarter way up the sides. Note that if larger quantities are required and a larger container is used, line the container with bacon as I have just described; the process is the same.

Place in a pre-heated oven at 350°F and cook for 1½ hours. Remove and you will find that the sides have shrunken in. Ladle some of the water from the tin into each container to fill, then leave to cool. When the pâté has cooled thoroughly, place the lids on the tops and label. If you want to make two pâtés, you can use smoked streaky bacon, remembering to remove the rind and label correctly.

MAYNARD'S TIP

With this mixture, sherry or brandy may be added according to taste. If you decide to use brandy, this increases the shelf life as alcohol is a natural preservative and also improves the flavour.

Beef Burgers

A very useful line, and ideal for the barbecue season. If you produce a good quality burger using good quality meat, and making sure that the meat is free of gristle and bone, this will increase your trade.

There is nothing worse when eating a burger than finding gristle and bone in the thing. You can sell anything once — the secret is selling it twice!

Ingredients

35lb good quality beef
7lb suet, with skin removed
1lb onions
6lb rusk
4 eggs
seasoning (see blue box)

Seasoning for Beef Burgers

Ingredients

15lb fine salt
2oz cayenne pepper
4oz ginger
2oz nutmeg
3oz mace

Use ½oz seasoning to every 1lb of meat.

Method

Put the lean beef and the suet through a medium grid. Mince the onion on a fine grid. Put the beef, suet and onion in the bowl chopper, add the seasoning and mix thoroughly; then add the rusk. If the mixture is too tight, add a small amount of boiled sterilised water.

Now whisk the eggs, add to the mixture, and turn the chopper until the ingredients have fully blended. Leave for 30 minutes so that the flavourings can permeate the mixture.

I recommend making the burgers by using large sausage casings. Using a straight filler (this is the sausage filler where the piston comes up from the bottom), attach a large nozzle to the horn. Use large synthetic casings. Tie one end and stuff out tightly — the casings can be stuffed tight as they are not going to be cooked in the boiler.

When stuffing, leave enough space at the end to tie off. When you have completed the filling, place the cased meats in a plastic tray, packing them tightly in the tray so they keep their shape.

Another method is to put the mixture in a press and press out individual burgers. I prefer the synthetic casing method because you can cut a large number; you put the filled casing on a flat-bed slicing machine and cut to the size you want. The meat does not fade or oxidise in synthetic casings, so another advantage is that you have some control over the product.

There are other ways of making burgers. You can buy a large automatic machine, which pumps them out and puts them on paper. Another method is to roll the mixture out on a table; you flour both sides and lay it out on greaseproof paper, then cut out burgers with a pastry cutter — this is a very time-consuming method and I suggest the first option is the best.

MAYNARD'S TIP

When cooking burgers or patties under the grill or in the oven, lightly coat with oil or butter. This prevents the outside drying out.

Spiced Beef

This is a lovely dish: sliced thinly with fresh bread and pickles, spiced beef will give you a meal fit for a king. But, like other recipes, spiced beef has often been spoilt by using very fat meat, predominately brisket and other fat cuts. Fat meat is not favoured by the public today — leanness is the key word. So, if you try this recipe, I suggest you use a very lean meat. I like to use shoulder of beef or similar lean cuts. In fact this dish can be a useful one for using up those good quality off-cuts and oddments that are the by-products of a busy practice and need to be used. But remember to use only the lean cuts of beef. Made well, this dish will bring you success.

Method

If you are starting with a substantial cut of lean beef you will need to follow the curing recipe (see Spice Brine recipe following). First, remove any fat, gristle and bones from the beef and place in a cleansing pickle for approximately 1 hour — this will remove blood and debris and start the osmosis.

Making the Spice Brine

Gather together the ingredients listed in the blue box (opposite page). Now put the water in a container and add the salt. Dissolve the sugar in a saucepan and add to the mixture. Grind the saltpetre finely, put in a dedicated saucepan with water, heat until dissolved, and when cool add to the mixture. Grind all the herbs and spices finely. Put in a saucepan with a small quantity of water, boil until infused, and when cool add to the mixture. The colour of this brine is important for colouring the beef, so I suggest you go to one of your established brines, take some out, boil it and add to the new mixture to give it some maturity.

Curing with the Spice Brine

The brine is now ready for the meat. It should have a density of 70%, but the density for pumping should be 40%, so take some brine from the container and

Ingredients for the Spice Brine

20 gallons of water (boiled and sterilised)

15lb rock salt

5lb sugar (do not use a dark sugar for it will darken the beef)

12oz saltpetre

1lb juniper berries

2oz cloves

1oz pimento

4oz ginger

8oz peppercorns

Lean cuts of topside and silverside beef are recommended for the spiced beef recipe

add some water to lower the density. Once 40% has been reached, this brine can be used for pumping. Do not over-pump the beef because you will destroy the tissue and the meat will break into small pieces when cooking.

When pumped, place the beef in the brine, which, remember, should have a density of 70%. Make sure all the meat is submerged and leave for 6 days. The meat should take on a reddish colour. Once the correct colour has been achieved, take out the meat, wash it off and put it in the boiler.

Fill the boiler three quarters full, and turn the temperature up to 160°F. Do not go above 160°F, as you will overcook and destroy the texture of the meat. If you wish, you can add a couple of spoonfuls of honey or a certain amount of malt vinegar (the vinegar brings out the taste). Cook until tender; the cooking time will depend on the cut and size of the beef.

When cooked, take out and put on a preparation table. You will need to trim the beef up so that it fits your square mould. The off-cuts will not be wasted and, as you will see, can be used to 'pack' the mould using the gelatine to bind it.

Rinse the moulds in the boiler so they become warm — it is best to use a square press for this beef as this is a better shape for slicing. Mix 2oz gelatine in a pint of water: use the water from the boiler as this will add flavour, but sieve before use. Once the gelatine has been mixed, pour a small amount in the bottom of the press; this makes a bed for the beef.

Now start to build the beef up. If you find you have several strips of beef, and perhaps also some off-cuts, I recommend you do this by going one way with the beef, then the other. When you have built up the beef in this fashion, and the press is snuggly filled, pour the gelatine over the meat. Leave for an hour and top up with gelatine. If you find the gelatine is too white, add some colouring that will blend in with the beef. Then put the lid on the press gently and leave to cool. It is most important to leave the gelatine to cool before you tighten the press. If you don't do this you risk squeezing the gelatine from between the strips of meat and they will fail to bind together. When cooled, lower the lid of the press down a notch; this will make the beef firmer for slicing. Now place the press in the fridge.

When the gelatine is set firm, remove from the fridge. Dip the press in hot water to loosen the meat. Turn out on a wooden board. Do not turn out on a stainless steel table as this may damage the press. An alternative method is to put a cotton cloth on the table. As the beef is released make sure it remains all in one piece! Wrap it in greaseproof paper and return to the fridge to mature — all cooked meats need time to mature.

MAYNARD'S TIP

Vacuum pack the spiced beef and this will increase the shelf life.

Sweet Cured Roast Beef

I think this is an excellent product; it takes a little bit longer to produce than the spiced beef but I think you will be rewarded with the result. At one time, this product was boiled. I was not a big believer in this method as I thought all the flavour and juices was lost from the meat. So I devised this method, and in my humble opinion I thought it was an improvement.

Method

The first thing to do is obtain some fresh topside or silverside of beef. Put in a cleansing pickle to clean for an hour or so. When clean, remove, dry thoroughly and vacuum pack. Place in the fridge for 7 days as this will help the beef to mature.

Make the sweet brine for this beef from the following:

25lb salt

10 gallons of water (boil to sterilise)

7oz saltpetre

2lb light sugar (I suggest light Muscovado or Demerara)

¼oz sodium nitrite

The density of the brine needs to be 70%. Sterilise a stainless steel bucket and dip this into the tank to remove some of the brine. Dilute with sterilised water to 40% density. Remove beef from the fridge and pump with the 40% density brine. It is important to try and pump through the artery which is at the top of the silverside; if this is not possible, pump gently with a straight needle. Do not blow them up, but pump them gently.

Test the brine for curing. This must be 70%. Stir the brine well before placing the beef in it. Once the beef is in the tank, put a non-metallic weight on top so the beef is below the water. The curing time is dependent on size. I suggest that if the beef consists of large pieces, they should be left for at least 8 days. The beef should have a pink colour, and not just on the outside; the colour should be all the way through the meat. Once you are satisfied that the pieces have the right colour and are sufficiently cured, remove from the tank — judging the time comes with experience.

Do not wash off, but wipe off to dry. Then place in the fridge for another week to gather flavour.

Now place the pieces on a stainless steel table — the next step is to lard them. Thread your larding needle with some of the fat you have cured by dry salting (see the 'Fat Bacon for Larding' recipe in Chapter 6). I used to coat strips of back fat with coriander and black pepper, and thread the strips of fat through the meat at approximately 4-6" intervals.

It is important to do this larding as when the meat is cooked it will be supple and have a lovely flavour. If we did not lard these pieces, they would be very dry as there is no fat in this beef.

Once larded, spread coriander, black pepper and allspice over the meat and then wrap in a cooking net. This is important as the net will keep the shape. Once the net is on, place the beef in a cooking bag, and place the bag on a baking tray. Place a small amount of salt in the bottom of the tray — this prevents the bag sticking to the tray.

Cook at 350°F; the cooking time is roughly 20 minutes per pound. When you feel it is cooked, remove from the oven and leave to cool in the bag for thirty minutes, then refrigerate. Once cold, remove from the bag and place on a rack to dry.

There are two methods with this product. One is to smoke the beef: if you decide to do this, I suggest that you lightly smoke it. Smoke using beech and straw. Place in a muslin bag and smoke at 110°F. Leave in the smoke house to cool down. This gives you two products: the smoked and the unsmoked. I think you will find the smoked one a better seller as the flavour improves with the smoking.

Roast Spiced Beef

Seasoning

    8oz coriander
    10oz black peppercorns
    3oz white pepper
    2oz allspice

Method

If you want you can add a small amount of garlic to the seasoning. Follow the procedures for preparing spiced beef. Cook this beef at 350°F. Cooking time is roughly 20 minutes to the pound.

Haggis

Haggis is a traditional Scottish dish that has become very popular in England and Wales in recent years. It can be made from the stomach of a calf, sheep, pig or lamb. It has a distinctive flavour and you can make it with a variety of ingredients, such as liver, tongues, beef and other oddments. In some places in Scotland, a sweet haggis is made using currants and raisins, but predominately it is a savoury dish. I think you will find that these recipes will go down well with your customers.

Scottish Haggis

Ingredients

    5lb sheep's liver, hearts and tongues
    1lb beef suet
    3lb oatmeal
    2 Spanish onions

Seasoning

    3oz fine salt
    1oz white pepper
    1½ pints of stock
    Juice of one lemon

Method

Put a sheep's or lamb's stomach in a strong solution of salt water to cleanse. Turn inside out and make sure it is very clean. After 30 minutes throw the water away and repeat the process — cleanliness must be paramount in these preparations. Clean the liver by placing it in a strong solution of salt water; this will remove any blood clots and debris. When clean, take out the liver and remove all the sinews; repeat this process with the tongue.

Boil the liver, tongues and other oddments. Keep the water from the boiler for your stock. When cooked, mince all the meat ingredients on a medium grid. Mince the suet on a fine grid; add a small amount of flour so that the suet does not stick to the grid.

Now put the oatmeal into a large container. Mix the oatmeal with the stock left from the cooked meat; mix into a fine paste and place in your bowl chopper.

Add the minced meats and finely minced onions. Add the suet and spread this evenly in the bowl chopper. Lastly, add the seasoning, spreading this evenly over the surface. Turn in the bowl chopper until the mix is of a fine consistency; add more seasoning if necessary until the desired taste is reached.

Check the stomach before stuffing to make sure it is well cleaned. Do not over-fill — fill three-quarters full to take into account expansion. Make sure the top is tied with a good knot; sometimes it is necessary to sew the stomach to close it.

Now place the haggis in the boiler and cook gently for 2½–3 hours, at no

The "honest, sonsie face" of the delicious haggis. A favourite of the Scots and, I am pleased to say, no longer underrated in England and Wales

hotter than 180°F. Any hotter and the haggis will burst. To check that they are adequately cooked, use a needle and prick one of them — the needle should come out clean if the haggis is well cooked.

It is best to remove the haggis from the boiler using a sieve. Place the haggis on trays, leaving a space between them to cool. Place in the fridge when cooled.

Some manufacturers put lights and rinds into haggis but I am not a big believer in that. I say, put the best in and get the best price.

Haggis 2

Ingredients

12lb pig's liver and tongues

4lb beef suet

4lb oatmeal

2 Spanish onions

Seasoning

4lb fine salt

2lb white pepper

1oz nutmeg

1oz cayenne pepper

Method

Follow the procedure described above. I think you will find that this variation will be a popular item on your menu, with a distinctive spicy flavour.

Potted Meat Spread

This has become a forgotten food. It has been badly made for many years and has lost its reputation, but I think it is time for someone to make a proper potted meat spread (sometimes known as sandwich spread). You could put it in china containers, in tin foil containers or in synthetic casings weighing about ½–¾lb — clip both ends of the casings and that will make them very attractive.

You could use a number of recipes for a spread, such as the pressed ham & tongue recipe above, and you could make different flavours at different times. The advantage of potted meat spread is that all the end-pieces of production can be turned into money, so there is an economic advantage to develop this product.

When making potted meat spread, I recommend you use a portion of cured meat as this will give the spread a longer shelf life and a lovely colour — most customers buy with their eyes.

Seasoning for Potted Meat Spread

Seasoning

4ozs white pepper

3oz Bay salt

1oz mace

¼oz cayenne pepper

Use approximately ½oz seasoning to every 1lb of meat.

Potted Meat Spread Recipe

Ingredients

14lb cured beef

7lb cured hocks or bacon ends & a small portion of belly pork

3lb butter or margarine

Seasoning (see recipe opposite)

Method

Weigh all the meat ingredients and place in the boiler. Cook at a steady heat but do not boil; cook until tender. Take meat out of the boiler and remove all the gristle, bones and debris. Put through the mincer on a fine grid.

Place the butter in a container. Take out some stock from the boiler and pour it over the butter. Now place the minced meat in your bowl chopper. Add the seasoning and mix well; then add the stock and butter. Keep turning the bowl chopper until you have a fine paste.

Test for taste as this is a highly seasoned product; this meat needs to be highly seasoned as it is eaten with bread and needs a distinctive flavour.

Note that the seasoning (see opposite page) has to be added according to taste, as precise amounts are hard to specify when cured meat is used in the recipe.

Have your containers ready. Scoop the potted meat spread into them, evenly distributing the paste. Leave to cool, then melt some butter and pour it over the top of the spread; this will seal the top.

Beef, Bacon & Tongue Potted Meat Spread

Ingredients

9lb cured beef

4lb bacon hocks or bacon fat ends

3lb tongue (lamb, pork or ox)

2lb butter or margarine

Method

Follow the procedure as described for Potted Meat Spread.

Seasoning for Beef, Bacon & Tongue Potted Meat Spread

Seasoning

6oz white pepper

½oz pimento

½oz cayenne pepper

salt (the amount will depend on the amount of cured meat you use in the recipe)

Use approximately ½oz seasoning to every 1lb of meat.

MAYNARD'S TIP

As alternative seasonings for the potted meat spreads, consider adding a little of any of the following to the dry mix: cinnamon, ground ginger or nutmeg.

Smoked Potted Meat Spread

Ingredients

>   2lb smoked belly, with rind removed
>   8lb shoulder pork
>   12lb bacon hocks, collar pieces and ham hocks
>   3lb butter or margarine
>   seasoning (see blue box)

Method

Follow the procedure as described above.

Seasoning for Smoked Potted Meat Spread

Seasoning

>   5lb white pepper
>   4oz mace
>   2oz ground ginger
>   1oz cayenne pepper
>   9lb Bay salt (finely ground)

Use approximately ½oz seasoning to every 1lb of meat.

MAYNARD'S TIP

You will need a bucket of very hot water for your knives; I recommend you have 3 knives in the bucket. Keep the water in the bucket hot at all times; this helps the knives to remain hot, thus making the cutting of the fat easier.

Scottish White Pudding

Ingredients

>   10lb barley or groats
>   10lb leaf fat or pork fat
>   3lb leeks
>   5lb farina
>   2lb milk powder
>   seasoning (see blue box)

Method

Boil the barley or groats in a linen bag for 40 minutes or so until swollen. Meanwhile, chop the leeks finely. Mix together the farina and milk powder.

Blanch the fat in the boiler at about 180–200°F; then cut into small pieces. Put the barley in a container and add the fat. Add the leeks, farina and milk powder and

---

Seasoning for Scottish White Pudding

Seasoning

>   1lb salt, finely ground
>   6oz white pepper, finely ground
>   2oz coriander
>   1oz pimento

Use approximately ½oz seasoning to every 1lb of meat.

mix until you have a smooth paste; if the mixture is too stiff, add some water.

Transfer to your bowl chopper and turn until the mixture is evenly distributed. Stuff out (but not too tightly) using beef runners or synthetic casings.

Cook in the boiler at 180°F for 30-40 minutes. Take out, leave to cool and you now have a Scottish White Pudding.

Breakfast Slice

This slicing sausage was predominately a northern dish, fried for breakfast and sometimes served between two rounds of toast. It was cheaper than bacon to produce but I can assure you it was delicious and a very simple thing to make. This is another forgotten recipe and I think you should know how it was produced.

Ingredients

40lb pigs' heads and hocks, lightly cured
10lb rinds, lightly cured
10lb pork liver
20lb oatmeal
20lb stock

Method

Place the pigs' heads, hocks and rinds in the boiler. Put the liver in a strong solution of salt to remove all blood and debris. Once cleaned, remove all sinews and place liver in the boiler.

Now boil the meat until tender; keep

Seasoning for the Breakfast Slice

Ingredients

6lb 6oz ground white pepper
4oz sage
2oz mace
1oz ground nutmeg
salt according to taste (there may already be enough salt in the cured rinds and head meat)

Use approximately ½oz seasoning to every 1lb of meat.

the water for your stock but sieve before use. Take out the meat and remove all bones and gristle.

Put meat through the mincer on a large grid, and place in the bowl chopper when this is done.

Turn twice in the bowl chopper. Before adding the seasoning, taste the meat for salt content and modify the seasoning accordingly. Now add the seasoning and turn twice.

The next step is to add the oatmeal. Add some stock and mix until you have the desired consistency. The mixture is now ready to put in the sausage filler.

Stuff into large-sized synthetic casings. Tie one end with cotton string, then fill. Wash off and place in the boiler.

Cook at 180°F for 1½ hours. Take out and wash under cold running water. Hang up to dry and leave to mature. After a couple of days, they are ready for use.

Garlic & Ham Sausage

Ingredients

15lb lean shoulder pork, cured

6lb ham pieces, cured

4lb medium-grade rusk

2lb farina

seasoning (see blue box)

Method

Mince the shoulder pork on a medium grid; do the same with the ham pieces. Put the farina in a container, add some water, and mix to a smooth paste. Put the paste and the meat in the bowl chopper. Chop until evenly distributed.

Before adding the seasoning, taste the mixture for salt content as cured meat is being used; modify the seasoning accordingly. Do not put whole garlic in the bowl chopper but put in a garlic press and crush first before adding to the seasoning. Add the seasoning and mix thoroughly. Last of all, add the rusk. If the mixture is too solid, add a small quantity of water to loosen. Mix to the consistency required.

Take out and rest the mixture, then stuff into synthetic casings. Do not stuff too tightly as they will burst in the boiler. Tie off both ends. Put in the boiler and cook to an internal temperature of 75°C (180°F) for 1 hour. Take out, wash off with cold water, then place in a cold bath. Fetch out when cold, dry off and place on sausage sticks. Hang in the fridge, and you now have a Garlic & Ham sausage.

Seasoning for Garlic & Ham Sausage

Ingredients

6lb Bay salt

2lb white pepper

2oz mace

3oz coriander

2oz pimento

3oz fresh garlic

As a rough guide, use ½oz of seasoning to 1lb of meat

MAYNARD'S TIP

It is best to make this recipe the last one of the day as the garlic will taint other ingredients in the bowl chopper.

Garlic sausage and two types of salami

Black Pudding

A famous English sausage, except that it is not of English origin by any means. The ancient Greeks had a form of black pudding, and the Germans, French, Italians and Americans had a similar product. The landed gentry as well as farm labourers enjoyed black pudding in England.

Some hunting fraternities would serve this on a hunt with a glass of sherry or wine and it went down very well. The farming community would eat boiled black pudding with mashed potatoes and peas and if any of you want to try that, you will have a marvellous feast.

Black pudding was made in all areas of northern England — especially in Lancashire, the Manchester area, and Cheshire — and to a lesser extent in southern England. Each region had its own distinctive recipe.

In North Staffordshire, black pudding was fried, put into an oatcake with bacon

Polony (red synthetic casing) and black pudding (natural casing)

211

Black pudding showing the large pieces of pork fat which characterise this traditional English product

and cheese, and then rolled up — this was the Staffordshire breakfast.

Black pudding is a traditional dish of England, made in different ways in different areas. I think that makes it interesting, but many people in the food industry today have finished making black pudding as they think it is a hard task and a messy business. That should not be the case if it is made correctly, and in this recipe I will explain how it is done.

If ever there was a product that needed to be produced, this is it. There is a ready market for it, and I think this is a golden oldie that has been forgotten.

Black pudding can be made using beef bungs (large beef casings) or middle and large hog casings; you can also use synthetic casings.

Horseshoe versus Straight

When I worked for old Theo, we used to make two types. We used to make a ringed version (shaped like a horseshoe) and a straight version — the original black pudding was made straight and it is easier to make them this way.

Then we had the development of the synthetic casing, so we had three ways of making black pudding: the horse-shoe version, the straight version, and the synthetic version. I will give you a selection of original recipes and the choice is yours.

Regulations over fresh blood

The main ingredient of black pudding is blood. If using fresh, as opposed to dried blood, the blood must be collected using a vacuum-knife system at the abattoir, under veterinary supervision. If you are using dried blood the conversion rate is 1 part dried blood to 5 parts water.

Note that some of the quantities of blood are given in pounds rather than pints, as this is how we used to measure small amounts of liquid in the industry.

Leaf or Flare Fat

To make a black pudding, you will need to use leaf fat from the pig — leaf fat is the fat from the belly. Leaf fat needs to have the tough surface membrane removed before it is diced. If leaf fat is not available, back fat is an alternative.

The fat needs to be cut into ¾" squares, and the best way to do this is to take the skin off the fat first. When you have sufficient squares of fat, put them in a muslin bag and scald at 200°F — this process seals them.

Binders

Black pudding generally includes groats (oats are my preference), which have binding qualities. Put them in a linen bag, leaving room for expansion as they cook, and cook at 200°F for 2 hours. If you prefer, you can use barley — in which case put it in a linen bag and cook for 3 hours, or until soft.

You can add other binders such as flour, farina or rusk to the ingredients of a black pudding and this will make a firmer mix. You can also add minced rinds, which have binding qualities as well as flavour. I recommend you do this as the rinds give the black pudding body.

On the following pages I will give you four recipes for black pudding, and I think you will find them all very tasty.

Black and white puddings showing the contrasting texture of the insides

Staffordshire Black Pudding

This black pudding was taken to work by the pottery workers, the miners and the steel workers of North Staffordshire. It was a very rich black pudding and gave sustenance to the people doing a heavy job.

Ingredients

20lb fresh pig's blood (or the equivalent volume made up of dried blood)

20lb leaf (or back) fat

5lb groats or barley

2lb medium-grade rusk

2lb oatmeal

2lb Spanish onions

seasoning (see blue box)

Method

Prepare the fat and the groats (oats or barley) as described above. Peel the onions and chop finely. Put the groats in a large container, followed by the fat, then the seasoning, then the onions — mix well.

My old method involved warming the blood, then putting it through a fine sieve to remove clots and debris. If you are using dried blood you must warm up the dried blood mix and then add it to the barley/onion mix. Now add the oatmeal, mixing very slowly. The last thing to add is the binder. This recipe uses rusk, but some Staffordshire black pudding makers use minced rinds, or a combination of rusk and rinds — the choice is yours.

Once your ingredients are well mixed, you are now ready to fill the casings. Cut your hog runners into yard lengths. Tie

one end (cotton twine) and — using the straight filler or the large (black pudding) filler — stuff the casings, but not too tightly. Make sure the fat is evenly distributed and tie off the end.

You can make these puddings into rings or long pieces; I prefer to make them into long pieces. Wash off in cold water and they are ready for cooking. Make sure the outside is clean; if any debris is left on the outside, the puddings will not look appetising when cooked.

Fill the boiler three-quarters full. Heat to 180°F. Place the puddings in the water and be sure to maintain this temperature. If the temperature drops off there is a risk that the pudding will not cook right through. The time it needs to cook depends a bit on the thickness of the casing.

Black pudding dye used to be applied to make the pudding especially black but this process is now thought to be carcinogenic and it is therefore banned. Another old-fashioned trick was to put a nugget of soda in the boiler 5-10 minutes prior to

Seasoning for Staffordshire Black Pudding

Ingredients
8lb fine salt

2lb white pepper, finely ground

2oz pimento

1oz cloves

2oz coriander

Use 6oz seasoning to 14lb of meat.

the end of cooking. It was important that it was given just 5-10 minutes cooking time because if you put the soda in too early, the casings had a habit of bursting. Again, the use of soda in black pudding production is no longer legal.

This black pudding takes about 30 minutes to cook. To test, take a pin and prick the skin. If no blood appears, they are cooked. I know it is old-fashioned today, but we would use a walking stick to hook the puddings out of the boiler. When cooked, take out, place on sausage sticks and leave to cool.

To give your black pudding an extra shine, rub it with some lard or olive oil once cooled. Note that black pudding must always be completely cold before placing it in the fridge.

Lancashire Black Pudding

Note that the quantities in this recipe are for large-scale production.

Ingredients

> 5 gallons of fresh pig's blood (or the equivalent made up of died blood)
> 30lb back (or leaf) fat
> 6lb rinds (well cooked and minced)
> 16lb groats or barley
> 10lb rusk (superfine grade)
> 10lb flour
> 1lb kibbled Spanish onions
> seasoning (see blue box)

Method

Follow the procedures described above for Staffordshire Black Pudding. Note that this recipe uses flour, rusk and groats, all of which act as binders, so the recipe does not require oatmeal (which also has binding qualities).

Seasoning for Lancashire Black Pudding

Ingredients

> 1lb fine salt
> 4oz white pepper
> 1oz sage
> 1oz marjoram
> 1oz thyme
> ½oz pennyroyal
> ½oz powdered celery

Apply about 6oz seasoning to 14lb of blood/caul fat.

MAYNARD'S TIP

Blanch the pieces of back fat in salted boiling water for 5 minutes. This will make it easier the dice and it will give it that attractive pure white colour.

Yorkshire Black Pudding

Ingredients

- 10lb fresh pig's blood (or the equivalent volume made up of died blood)
- 4lb back fat
- 5lb minced rinds and head meat
- 3lb groats or barley
- 1lb medium-grade rusk
- 1lb oatmeal
- 8oz Spanish onions

Method

Put the rinds and head meat in the boiler and cook at 180°F until tender. Take out and leave to cool; then remove all the bones and gristle, and put through the mincer on a medium grid. Put the fat into a cooking net and cook until blanched; take out and leave to cool. Run the blood through a sieve to remove clots and debris; then put the rinds, head meat, and fat into a tub with the blood — mix well.

Seasoning for Yorkshire Black Pudding

Ingredients

- 7oz salt
- 4oz white pepper
- 1oz allspice
- ½oz cloves

Use about 6oz seasoning to 14lb of blood/fat/meat.

Prepare the groats (or barley) as described above. Peel the onions and chop finely. Add the chopped onions to the mixture, then the groats, then the seasoning, then the oatmeal, mixing all the time. Finally add the rusk; this will tighten the mix.

Stuff out into synthetic casings, leaving room for expansion. Cook for 20-30 minutes in the boiler at 180°F. Take out, wash off, and leave to cool. Put in the fridge to mature. After a day, they are ready for sale.

Baked Black Pudding

This black pudding is baked in the oven and was popular in hotels, cafés and the catering industry.

Ingredients

- 16lb fresh blood (or the equivalent volume made up of died blood)
- 8lb beef suet
- 3lb barley
- 1lb plain flour
- 2lb oatmeal
- 2 onions
- 2lb rice
- seasoning (see blue box)

Method

Cook the barley until soft. Cook the rice likewise. Put the barley and rice in a large container, then add the blood. Mix well; then add the flour and mix until evenly distributed. Chop onions finely. Put the

Seasoning for Baked Black Pudding

Ingredients

    7oz salt
    4oz white pepper
    1oz allspice
    1oz ground coriander

For the seasoning, use about 6oz to 14lb of black pudding ingredients.

suet through a fine grid. Add the chopped onions to the mixture, then the suet, then the seasoning, then the oatmeal, mixing all the time. If you think the mix is not tight enough, add more flour and mix thoroughly. Grease your baking tins and place the mixture in the tins, covering them with foil. Bake in a moderate oven. When cooked, fetch out and leave to cool. This pudding can be sliced and fried with egg and bacon or a mixed grill.

Natural Lard

All the fat from the factory can be siphoned into this natural product. Lard can be put in small containers for general sale to the public, or put in large containers for pie production. Using your own fat in the production of your pies and pastry gives your pies a better flavour, and I think it is a worthwhile operation as you are using everything and wasting nothing.

Method

Put all of the fat through the mincer, using a fine grid. Put half a gallon of water in the boiler and light; use a moderate heat. Add small quantities of minced fat; wait until dissolved then add more until you have used all of the fat. Stir all the time the lard is cooking. Once it is all melted, add some salt to preserve the lard and to stop the fat sticking to the bottom of the boiler and burning (add about 6oz salt to 30–40lb of fat).

You will need two boilers for this process. When the lard is well cooked, sieve it into a second boiler, using a fine sieve; this removes any residue. Put this residue in a lard press (the residue is known as greaves). Greaves would make great dog biscuits.

Stirring the lard in the second boiler, add some bicarbonate of soda, roughly ½oz — this will help the colour. A scum will appear on the top of the lard. Using a sieve, ladle off until the lard is clear, and dispose of the scum.

When you are satisfied that the lard is well cooked, there are two methods of

A block of lard

presentation: large blocks or smaller ones. For large ones, put the lard into square moulds. Put a polythene bag in the mould and pour the lard into that; then you will find that when the lard is set all you have to do is to tip the press upside down and removal will be easy. Date the lard and store in the fridge.

If you do not want to use large containers, add a small quantity of rosemary and put the lard in small containers — these are more suitable for sale to the public.

Pork Scratchings

A famous dish of the Midlands. Pork scratchings are the rinds, which are your final leftovers once the fat has been extracted for producing lard. The rinds are put into square trays in a moderate oven and baked until they are hard. Then they are taken out, left to cool, and coated with a thick layer of sea salt. They are put on trays, with a space left between them, and packed into bags, each one weighing approximately ½lb. Try this, and you will find you have quite a delicacy.

Pork scratchings, the perfect accompaniment to a pint of beer!

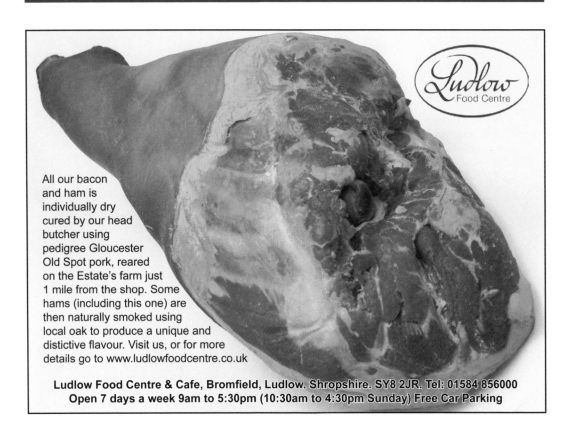

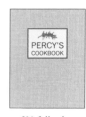

GLOSSARY

BASTING MIXTURE
A mixture of ingredients to enhance flavour, ladled over meat during cooking.

BED OF SALT
A 3-4" layer of salt, on which to lie the meat.

BEEF BUNGS
Large diameter sausage casings, made from the intestines of beef cattle, used for making black puddings.

BINDERS
Binders are used to trap meat juices, bond ingredients, and absorb the water in a product; they include rice, rusk, farina, milk powder, and flours of different kinds.

BLOWTORCH
This is a special blowtorch, used by butchers to remove hairs from the pig; also used by chefs.

BOAR
A male pig, not castrated.

BREADCRUMBS
Dried bread, cooked and crushed; used to enhance the appearance of cooked hams.

BRINE TANK
A large container used for holding a brine, made from plastic or stainless steel.

BRINE
A salt solution for curing meat.

BRINEOMETER
A device that measures the amount of salt in a brine, expressed as a percentage: a brine with a 70% density has more salt than a brine with a 40% density.

CAUL (or KELL) FAT
The thin layer of fat that surrounds the abdominal organs.

CLEANSING PICKLE
A mixture of salt, saltpetre, and water, used to clean meat and start the curing process.

COOKING NET
A cotton net used in cooking to put small items in when boiling them, or to wrap around meat to keep its shape.

CURE
A dry cure is a mixture of salt and saltpetre (and other ingredients to add flavour); a wet cure is a solution of salt and saltpetre (and flavourings) and is also known as a pickling solution or a brine.

CURED MEAT
Meat that has been treated with salt and saltpetre: the salt dehydrates the meat, and the saltpetre gives the meat a pinkish colour.

EQUALISE
The process after curing in which the salt, nitrates and sugars spread evenly through the meat; this takes varying periods of time according to the size of the meat.

Glossary

FORWARD
The term for cured or uncured meat that is old or running out of 'shelf-life'.

GILT
A female pig that has not had a litter.

HYGROMETER
Device for measuring humidity inside a curing room.

KELL (or CAUL) FAT
The thin layer of fat that surrounds the pig's abdominal organs.

LARDING NEEDLE
A needle with a large eye, used to thread fat into meat that has little fat content.

LARDING
The process of inserting long thin pieces of fat into meat.

LARDONS
Chunks of cured fat, used to flavour other products.

LEAF (or FLARE) FAT
A layer of fat that covers the pig's belly.

MATURE
The stage when the hams are hung (in about 40°F) and the white bloom develops. The cure has then been evenly distributed.

MEAT HAMMER
A wooden or stainless steel tool used to flatten meat.

MUSLIN CLOTH
A cloth made from cotton, used to protect a product from flies and bacteria.

NATURAL COLOURING
A natural ingredient used to add colour to your product, such as beetroot.

OSMOSIS
The process of removing fluid from meat by applying salt around it.

OVERHAULING
Moving the meat around in the curing brine to ensure that no part is left untouched by the brine.

OXIDISE
Combine with oxygen: when meat is left too long in the open air, the colour goes darker.

PEA FLOUR
Flour made from ground and dried peas, used to enhance the colour of bacon in the smoking process.

RIND
The skin of the pig.

SALTPETRE
Potassium nitrate, a chemical that prevents the growth of bacteria and when converted to nitrite it gives the pinkish colour to meat.

SAUSAGE STICK
A long stainless steel stick used to hang sausages for cooling purposes.

STERILISED WATER
Water boiled to 212°F, then cooled and covered.

STITCH
'Stitch' or 'stitch pumping' is the technique of pumping brine into meat using a brine pump and a special needle known as a brining needle. A brining needle has several holes along the side of the shaft to facilitate distribution of the brine into the meat.

STUFFING OUT
A term for putting meat mixtures into sausage casings.

SUET FAT
Beef fat from around the kidneys.

TROTTERS
The feet of the pig.

VACUUM PACK
A bag that has had the air removed out of it, which helps to preserve the product.

APPENDIX

TABLE OF BRINE STRENGTHS

The table below shows the approximate weight of salt required for different strengths of brine (taken at 60°F to give consistent readings). The strength, also known as density or salinity, can be measured with a brineometer (see left).

You should remember that all salts have slightly different strengths: the strongest I recall being Maltese sea salt.

The table below offers a rough guide to brine strengths, but you should always conduct your own tests with the brineometer.

% Solution	Imperial gallon	US gallon	Litre
40	1lb 3oz	1lb ¾oz	118g
50	1lb 8½oz	1lb 4¼oz	150g
60	1lb 14oz	1lb 9oz	185g
65	2lb 1oz	1lb 11½oz	206g
70	2lb 4¼oz	1lb 14oz	225g
80	2lb 10¾oz	2lb 3½oz	268g

BIBLIOGRAPHY

Bull, Sleeter. *Meat for the Table.* New York: McGraw-Hill. 1951.
Duncum, William John. *A Picture Guide to Dissection: With a glossary of terms used in the meat trade.* Modern Meat Marketing Series. London: Barrie & Rockliff. 1960.
Gerrard, Frank. *Sausage and Small Goods Production: A practical handbook on the manufacture of sausages and other meat-based products.* Fifth edition. London: Leonard Hill. 1969.
James, A. *Modern Pig-Keeping.* London: Cassell. First edition. 1952.
Kurlansky, Mark. *Salt: A World History.* New York: Penguin Books. 2003.
Livingston, A. D. *Cold-Smoking & Salt-Curing Meat, Fish & Game.* Guilford, Connecticut: Globe Pequot Press. 1995.

Healthy pigs at Acton Scott historic working farm near Church Stretton, Shropshire. (Open to the public: 01694 781306)

A new brood of splendid free range Large Black pigs at Percy's Hotel & Restaurant which forage freely in the estate's woodland at Virginstow, Devon.

INDEX